MEDICAL METAMORPHOSIS

The three step cure for America's health care crisis

J. R. WAGGONER, MD

Medical Metamorphosis

ISBN 978-1-4357-0182-3

DEDICATION

To Barb, Nate, Jamie, Mandy, and David, Liss, Hal, Max, Jess, Myrissa, and Shauna, Priss and Charles, and Burt. To the memory of my parents, Bob and Bev, and to H.D. Giedd and all our silly dreams.

You don't choose your family. They are God's gift to you, as you are to them. ~Desmond Tutu

Medical Metamorphosis

ACKNOWLEDGEMENTS

This book's message is important, and I will readily admit that its telling would not have occurred without the belief of a great many people—not in my skill as an author but in the message.

Every time I tossed the book in the garbage, someone came along and in one way or another pulled it out.

That process alone was enough to sustain my three years of work.

I believe this book has existed in 573 different versions. Every one of them was reviewed by someone strong-armed into service: family members, the Dewsweepers, fellow docs, an employee of the local Office Depot, Lucidins—an insurance agent with more common sense than all of the CEOs of all the companies he represents, a New England journalist I heard on NPR who was incredibly generous of his time, a buyer from e-bay and her learned friends, and a former Lieutenant Governor who offered me the gift of his wisdom. These people cared enough to be honest. I thank you all.

When I finally found an editor willing to take on the task of wading through the work of one more neophyte, I ended up with a former goalie. He must have been pretty good because after blocking all those 90 mile-per-hour slap shots, he still had a wealth of insight, patience, and expertise. Thanks Mister Butterfly.

Thanks to Rick for first showing me about Google.

There were also those who nursed my soul, boosted my spirits, and carried my backside. Ed H., you are truly the best. You did all three, graciously and with infinite understanding and generosity. Mitch, you gave me the courage to take the first step. JH, I'd never have survived long enough to start this book if you hadn't been my chief cat wrangler. Doctor D., you not only nursed my soul, you polished it. JKC, the book would have been done a lot sooner if I'd had you covering my back. Mother Earth, I'd hate to count the number of storms for which you were my anchor. And the graceful long-necked bird, your firm but gentle counsel is a jewel of many facets. Sheriff, the bad guys are in trouble because you got a new deputy. Father Ph, blessed are those whose eyes are open to the works of God, who patiently put up with the works of men, and know the difference between the two. And to the Stomes—I have never met a more generous, understanding person.

I bow deeply before you all in gratitude and love.

TABLE of CONTENTS

STEP TWO---
CREATE AN INFORMED CONSUMER WITH ONE CHANGE 93

Medical Metamorphosis

PREFACE

Since the Second World War, American medical care has focused on disease and not people. Then, thirty years ago, we embarked upon an experiment testing the proposition that a health care system run according to the tenets of an unfettered free market would decrease cost and increase quality.

That America should base its health care system on these two principles is not surprising. Our medical technology increased exponentially during the Second World War, and capitalism and free market philosophy are the muscle, bone, and sinew of America's strength.

But the sole purpose of medical care is to relieve pain, suffering, and disability in *people*, individual human beings. A human being is not a collection of organs that go awry. A human being is a unique biologic creation, an intact unit that thinks and feels—the single most complicated unity on the face of this earth. If you don't think so, just ask one.

And health care has none of the components needed for the economic experiment to succeed. A system ruled by supply and demand is made up of consumers buying things in a reasonably rational manner and vendors aggressively competing with each other to sell things like automobiles and laundry soap. As a commodity, medical care is unlike laundry soap or automobiles. Patients are far from typical or rational consumers.

And physicians cannot be twisted into simple vendors of services without wringing from them their capacity to be healers.

Now, after six decades of *disease oriented medicine* rather than *patient oriented medicine,* and three decades into the economic experiment, American health care costs twice that of any other industrialized society, and that care is not only impersonal, it is often dangerous and not even available to all Americans. A conservative assessment of health care's quality estimates that annually 100,000 Americans die premature deaths from medical errors or inaccessibility to medical care. There is no single statistic to support the claim that American health care is the world's best. America is the *only* industrialized nation that does not have universal health insurance.

The principles upon which we built our health care system have failed, but in spite of the loss of life, the exorbitant cost, and the dehumanizing care, the system staggers on.

In truth, the causes of the system's failure are simple. Further, the only effective solution to America's health care crisis is also simple. However, simple truths are sometimes obscured by the fog of self serving rhetoric.

The purpose of this book is to burn away that fog. It is an audacious proposition that three steps will cure America's health care crisis. It is an outline of a medical metamorphosis.

I am going to convince you—or at least *try* to convince you—that America's health care need not be the world's most expensive and still not serve its population. I am going to demonstrate how simple realities explain why our system has faltered. I am going to show the power of the doctor/patient relationship, how its dissolution has driven health care costs up and quality down. And finally, I am going to outline a structural change—one single change—that will liberate market forces so that they can work as they should—driving costs *down* and quality *up*.

I am not going to put you to sleep with complicated economics or systems theory. This book is derived from first hand experience and that will be the bedrock of its foundation. Read further if you're worried about the health care of yourself and loved ones. Read further if you have the sense that the complicated schemes being proposed are all sounding a bit like leftovers served up as a gourmet meal, and worse, maybe the leftover ham is slightly tainted. Read further if right now you are thinking, "This guy has *got* to be full of baloney."

I'm not.

Medical Metamorphosis

INTRODUCTION

Before venturing into solutions, let's take a brief look at the problems.

We will spend most of our time dealing with individual patients. That's how health care is delivered—one patient at a time. This is one of the simple realities ignored by most health care planners.

Health care is certainly *experienced* one patient at a time. Whatever you've got wrong with you is in truth unrelated to the entire rest of the world. Whatever you've got, you've got 100 % of the time. You may have something that only 1 % of the population ever gets, but if you've got it, it's all yours. That's what experiencing health care as an individual is all about. *No two patients are exactly the same.*

So, let's take a look at two *individual* circumstances that are typical of America's health care crisis.

Snapshot: The mother

Nancy's daughter pounds on the bathroom door, imploring her mother to hurry. Normally, Nancy would respond with maternal sarcasm, but today the forty-year-old mother of three stands silently, her eyes dilated with fear. She is sure she has just received a death sentence. Her right hand rests upon her left breast. She moves her fingers over a small, firm marble of tissue beneath the skin. Her own mother died at age forty-two, less than a year after she too had found a small marble of tissue, a lump, a seed of death called breast cancer. Nancy has feared this moment, but it is even worse than she had thought it would be.

She has schooled herself about this awful thing, this thing that might take her from her children. She has learned about early diagnosis, mammograms, and ultrasound-directed needle biopsies. She has faced the mathematics of five-year survival times. She has prepared as one would prepare for a life or death battle because in her heart Nancy knows that is the battle she will have to fight.

Many women would not have girded themselves for this battle, preferring to ignore its possibility. But Nancy has convinced herself that her outcome will be different from her mother's. Medicine has new weapons. She will fight the cancer because she wants to live, for herself, for her children. Her mother died seemingly without protest, completely overwhelmed. Nancy has been haunted by the thought that her mother did not love her enough to fight the battle.

Now, Nancy's preparation seems silly. She is defenseless. Some weeks earlier, her husband's company dropped medical benefits. She has no medical insurance. She and her husband have been trying to find some way of protecting themselves, but now it is too late. She knows that the aggressive kind of cancer that killed her mother will win the battle unless she begins fighting back immediately. She has no doctor to turn to. She cannot schedule the mammogram. There will be no needle biopsy. She is at the mercy of the marble of tissue that rests beneath her fingers. She closes her eyes as they fill with tears of hopelessness because cancer has no mercy.

Nancy lives in the richest, most powerful country in the world. She well may die because even though her country spends more on health care than any other country in the world, it has chosen to leave her and millions of others like her without the means of defending herself. How can this be?

Is this good health care?

Snapshot: The emergency room doctor

The patient listens intently as the emergency room doctor explains his problem. He is not having a heart attack. He is relieved by that news but also immediately embarrassed. He thinks it's pretty damn silly for an intelligent, tough vice-president of acquisition to go to an emergency room with esophageal reflux. "Reflux? Call it what it is," he says to himself, "heartburn." He almost did not go to the emergency room because he feared he was *not* having a heart attack. The last thing he needed was someone else telling him he had overreacted and that things weren't as serious as he feared.

The fresh-faced doctor had sensed that was how he might feel and with a consummate skill that is probably god given, presented her findings in such a way that the patient's ego was never challenged. That is why he listens as she describes not only how serious reflux might be, but also how with his family history

17

he was wise to worry about his heart. She performs her job at a level of skill unappreciated by even her peers. Had the he seen one of the other emergency room physicians, the patient probably would not have listened. Lacking the skill to intuit the patient's character, the time to talk to him, and the inclination to extend themselves in a setting where such extension is rarely rewarded emotionally and never rewarded financially, they would have treated the patient much differently.

But the young emergency room doctor is different. She defines her job as treating patients, not just defining the presence or absence of a particular disease. She realizes that unless she finds a way of actually communicating the information she has gleaned about the patient, the thousands of dollars spent obtaining that information will have been wasted. In her mind that is simply common sense. So, using the subtle skills of a healer, she finds a way of making him listen, a way of *allowing* him to listen.

This is the doctor's last shift as an emergency room physician. This patient may be the last patient she ever sees. After just a few short months as a practicing physician, she is leaving medicine. The profession she loves, the profession into which she has invested fifteen years of her life, has driven her from its ranks. She leaves not because she is incapable of performing under pressure, or because she lacks technical skills, or because she is poorly trained. She leaves because her profession demands that she make a choice between practicing medicine the way she knows it should be practiced and the way a hospital's financial goals dictate it must be practiced. She leaves because she has been told in no uncertain terms that the subtle skills of a healer are of little or no value.

When all the physicians like the young, talented emergency room doctor have left medicine, who will be left to care for America's patients?

Medical Metamorphosis

STEP ONE

RECOGNIZE WHY A SIMPLE SYSTEM OF SUPPLY AND DEMAND HAS FAILED HEALTH CARE

If you ask someone who worships slavishly at the alter of a free market why it has failed health care, you will be told rather curtly that in truth, it is the American public and the medical community that have failed the free market. You will be further lectured on the inevitable success of supply and demand in solving the health care crisis, provided that all parties begin behaving as they are supposed to.

"Patients need to be intelligent consumers," they say. "Doctors need to be vendors of services, competing with each other in a transparent and even aggressive fashion. And the government needs to keep its hands off the entire affair and let the chips fall where they may."

But what if it's not possible to strong arm doctors and patients into the roles of vendor and consumer? What if health care is not so much a typical commodity as a reaction to the natural disasters of illness and injury? What if it's like a boat during a flood or a piece of high ground during a tsunami?

Would such a boat or such a piece of real estate *ever* be valued according to supply and demand, or would their price remain horribly elevated until the waters receded?

The waters of illness and injury never recede.

Bentley's Second Law of Economics: The only thing more dangerous than an economist is an amateur economist.
Berta's Fundamental Law of Economic Rents: The only thing more dangerous than an amateur economist is a professional economist.

--Anonymous

CHAPTER ONE

THREE REASONS FOR HEALTH CARE'S ECONOMIC FAILURE

The experiment, a transformation into the "health care industry"

The economic philosophy of the doctors who trained me almost forty years ago was "Take care of your patients and the money will take care of itself." America's health care system was a reflection of that philosophy. In the late 1960s, virtually all hospitals and insurance companies were not-for-profit. The idea that a physician would advertise was heretical. A doctor might be censored by his local medical society for having a Yellow Pages ad that was more than the standard small-print name, address, and phone number.

As health care began to mirror the "miracle of modern medicine," its growth drew the attention of both the government and Wall Street. Because of Medicare and eventually Medicaid, the government was concerned about cost. Wall Street, on the other hand, saw an opportunity for profit.

Thus, it was decided that America's health care needed to be changed from a provincial system run by religious organizations and doctors into a modern *industry*, operating according to the principles of supply, demand, and competition. Consumers would then drive down price according to the natural effects of supply and demand, and competition would increase the quality of medical services. Similar arguments had been used in support of the deregulation of airlines and the breakup of the large telecommunications companies. The power of America's free market was going to be turned loose.

The traditional and conservative medical community was opposed to such change. Those of us just beginning practice were caught in the middle of a philosophical battle between our mentors, who at times could be somewhat self-righteous, and a business community, which at times could be overbearing and ignorant of the subtleties of caring for patients. The business community prevailed. In Denver, for example, all but one insurance company are now for-profit as are the overwhelming majority of its hospitals. Today, doctors have huge ads in the Yellow Pages, billboards throughout the city hawking their talents, and commercials on radio and television.

My medical practice was on the campus of Denver's first for-profit hospital. My office opened its doors at almost the exact same time as the hospital. I observed first-hand the birth, adolescence, and maturation of the medical industry. In fact, because I helped form the medical staff of just the second metropolitan hospital owned by Humana, Inc. (the corporation that would eventually become part of Hospital Corporation of America [HCA]), I had a unique opportunity to participate in this process as well as observe it.

It took me a long time to come to the realization that what was being attempted wasn't working. It wasn't until I had spent months studying health care policy and economics that I understood why.

Ok, if we just squeeze this and stretch that...

There is an old joke about a fellow who goes into a clothing store looking for a suit. He finds one he likes but it is not in his size. A salesman walks up to him and says, "That would be a great color on you, sir. Why don't you try it on?"

The fellow shakes his head. "Not my size."

The salesman takes the suit off the rack. "Look," he says, "expensive suits like this run slightly large. I'll bet this will fit fine."

"Actually," says the customer, "that would make it worse because I take a size smaller than that."

"Did I say they run large?" says the salesman. "Forgive me. Haven't had my coffee yet. I meant they run small." He shoves the suit into the fellow's hand and leads him to a dressing room. "Just try it on. If it doesn't fit, we'll look at something else."

Two minutes later, the guy comes out wearing the suit. He's swimming in it. "Wow!" says the salesman. "It's a perfect fit."

The guy stumbles over a pants cuff dragging on the floor and almost falls down. "Are you crazy?" he says. "This thing's so big that I gotta take two steps before my pants start heading in the same direction as my legs."

"Don't be silly," says the salesman. "Look. Grab the front of both legs of the pants and pull up on the pleats about four inches."

To humor the salesman, the fellow pulls up on his pants.

"See?" says the salesman. "Look where your cuffs break across your shoes. Your pants are a perfect length."

The guy looks down. Sure enough, his pants are a perfect length. "But look, buddy," he says to the salesman, "if I hold my pants up like this, my shoulders are all hunched up and the coat bunches around my neck."

"Don't be silly." The salesman pulls down both sides of the coat and pushes in the guy's elbows so they hold down the coat. "See? Doesn't that feel better?"

"But now I'm lopsided," says the guy. "This side is a little higher than the other and..."

The salesman doesn't even let the fellow finish his complaint. He takes the guy's right thumb and hooks it in the button hole of the left side of the suit coat. At the same time, he grabs the guy's arm and walks him to a mirror. Of course, since the guy is holding in and down the sides of the coat with his elbows, holding up his pants with his hands, and adjusting the left side of the coat with his right thumb, he doesn't so much walk as lurch.

Once the salesman gets the fellow to the mirror, he turns him one quarter towards a left profile and remarks in a low voice, "My goodness. I've never seen a more Savile Row piece of tailoring."

"Tailoring?" exclaims the fellow. "This isn't tailoring. This is me holding up my pants, and..."

The salesman interrupts the fellow, "Wait, wait, wait. Look in the mirror. Look. Just take a look at yourself. I'll be back in a minute."

The fellow takes a look, and sure enough, the suit appears to fit. After a moment, he realizes that its color really is the color he was looking for. "Perhaps," he thinks, "they might have it in my size ..."

The salesman returns, pulls one elbow slightly away from the fellow's side and slides a sack under it. Then he smiles and places the man's wallet, which he must have taken from his pants, into the coat pocket of the suit. Then he begins ushering the fellow towards the door.

"What are you doing?" asks the man. "What's in the sack?"

"Why your other clothes, sir," says the salesman. "I've packed them away, all folded nicely in the sack, and I've helped you along by charging your suit on your credit card, returning your card to your wallet, and, as you saw, placing your wallet in the coat pocket of your truly well-fitted new suit."

They reach the door and the salesman gives the fellow a gentle shove outside. It's gentle, but it's still a shove. The man is absolutely stunned. He starts walking towards his car—or rather lurching towards his car, adding a slight stumble every few feet, holding up his pants, down, in, and over his coat, with a sack under one arm.

Two doctors walk by and regard him with great sympathy. After they have passed him by two or three strides, one says to the other, "Good heavens. What horrible thing do you think befell that fellow, a stroke?"

"Perhaps," says the second doctor, "or maybe a spinal cord injury. Have you ever seen a more distorted gait?"

"Never," says the first doctor. "But I'll tell you what. That suit of his? What a well-fitted suit. I wonder who his tailor is?"

The attempts to fit America into a medical industry are not unlike the salesman fitting the poor fellow into the suit. Those who created the medical industry have asked patients to hold up their pants so they don't trip over their cuffs and instructed doctors to tuck in their elbows so the suit coat doesn't ride high and look lopsided. In truth, the suit doesn't fit.

Let me change hats from the crushed chapeau of an amateur comedian to the top hat of an amateur economist. Let's take a look at a free market system ruled by supply and demand to see why the suit doesn't fit.

Such a system has three components: a commodity or service to be sold; consumers who purchase that commodity or service, the demand part of the system; and vendors of that commodity or service, the supply part of the system.

Let's consider widgets as a commodity (a widget is a mythical thing that exists only in the mind of economists). Theoretically, the more widgets that are produced, the greater will be their supply, and the cheaper will be their cost for buyers of widgets (consumers). However, for this to be true, there must be more than one widget maker (vendors) because for the cost of widgets to go down, there must be competition between widget makers causing them to drop the price of widgets so that they are not stuck with lots of inventory. The other way for a maker of widgets to avoid unsold widgets is to make a better widget. Then, they might even be able to sell all their widgets for a price greater than those made by competing widget makers.

Turning from widgets to health care, the theory of market efficiency proposes that creation of a health care *industry* will force physicians (and all suppliers of health care services) to compete as vendors. Medical care will decrease in cost while increasing in quality because consumers will seek high quality care at the cheapest price—just like what happened with widgets. But unlike widgets, with health care, the opposite happened. Cost of health care has increased while quality has decreased.

Why?

There are three simple reasons that an unfettered free market health care system has failed to behave as predicted:

1. Health care is not laundry soap.
2. Patients are not typical consumers.
3. Physicians cannot be vendors; they must be professionals.

In other words, *a health care system has none of the elements necessary for a successful free market system of supply and demand.* For thirty years, economists, health care planners, politicians, and academics have grunted and groaned trying to perform an act of alchemy and change health care into laundry soap, trying to bend patients into consumers, and trying to hammerlock physicians into vendors. They have failed. The suit does not fit.

Unfortunately, during that same period of time, a number of tailors have made a great deal of money selling suits that don't fit, pushing us all out the door into the street holding up our pants while trying not to fall down. The American health care system has made some people very rich. Those people have no intention of simply standing by and allowing the thirty-year experiment to come to a close.

Thus, patients are chastised for not taking "individual responsibility" and becoming "educated health care consumers"; physicians are told they have not been "competing properly" and accused of violating public trust by inappropriately creating a demand for unnecessary medical services; and the true nature of medical care has been obscured in an attempt to ignore its life-or-death implications. And the experiment continues, and some people become very, *very* rich.

It's not rocket science

The basis of my three reasons for health care's failure is not complicated—it's not rocket science.

Consider # 1, "Health care is not laundry soap." If you were the CEO of Proctor and Gamble™, how would you like to be able to run an ad that claimed, "... finally folks, if you don't use the new improved version of nuclear-powered Sudso-Rinso-Cleano-Bright, *you may very well die!*"? Run that ad with conviction a few times, and how much Sudso-Rinso-Cleano-Bright do you think you'd sell? My guess is that even grungy college students would be lining up to buy the soap. The threat of death is a fairly strong selling point.

Of course no one would make such a claim. It's silly. A marketing strategy that even came close to such a claim might be a career builder for America's next Eliot Spitzer, and the ad

31

company for Sudso-Rinso-Cleano-Bright might have a few of its executives using their own product in the laundry of a federal prison. But an emergency room doctor who walks up to the parent of a child writhing in pain on an emergency room cart can honestly say, "I think your child has an infected appendix that may soon rupture. If she doesn't have an appendectomy, she could rupture her appendix, develop peritonitis, subsequent sepsis, and die."

If one regards this as a sales pitch and the doctor's medical advice as an attempt to "sell" a service, then it's about as powerful a sales pitch as one will ever encounter. It has life or death leverage.

Medical care *is* a service, a commodity, with life or death leverage. Laundry soap is not. "Neoclassical economics is built on the assumption that humans are rational beings who have a clear idea of their best interests and strive to extract maximum benefit ...from any situation."[1] There are many circumstances that can affect just exactly how rational a consumer is, but none more so than illness or injury in oneself or a loved one.

Even choosing a medication, which is a less emotionally charged situation than a sick child, is different from choosing laundry soap. If a doctor says to a patient, "We have a number of options, but the best one is 'X'," how often will a patient choose a different medication because of price if the consequences of

---

[1] "The triumph of unreason?" *The Economist*, January 13th-19th, 2007

that choice are stroke instead of no stroke, heart attack instead of no heart attack, or pain instead of no pain?

Ring around the collar is not a life-or-death matter. Health care is not laundry soap.

The truth of reason #2, "Patients are not typical consumers," is equally obvious. I once had a patient who was a local politician. His signature issue was the cost of health care and how the system was over utilized by patients and doctors. His contention was that patients sought care for circumstances that were inappropriate and unnecessary. Further, he contended that doctors paid no heed to cost or, worse, performed unnecessary services to line their pockets.

We had discussed these issues but only at a superficial level. I have a personal rule against debating patients, particularly with regard to politics or religion. As a physician, my responsibility is to understand what patients believe, not try to change their minds.

One evening I was paged on my way out of the hospital. This patient, whom we will call Jim, was in the emergency room with what appeared to be a kidney stone. I decided to see him since I was still in the hospital, and often times a familiar face, particularly your doctor's, makes the emergency room less threatening.

I heard Jim before I saw him. The way people respond to pain is a product of culture, personality, and circumstance. Jim was moaning and swearing at the same time. I pulled back the curtain of his cubicle, and when he saw me, he said, "Oh thank God. Thank God. Doc, help me. I'm dying. I can't take any more of this."

I walked to his bedside, and he grabbed my hand with both of his just as another wave of pain washed over him. His face contorted and for a moment I feared I would become a patient myself with a hand fracture.

I had already seen his chart, and he indeed looked as though he had a kidney stone. His urine was loaded with red blood cells, and the history he had given the nurse was very typical. He'd not yet been given any pain meds, and I ordered them. I hastened their administration by following his nurse as if I were some sort of stalker. The medication cut his pain—when a kidney stone is at its worse, no amount of pain medication will completely *eliminate* its pain.

Remembering his philosophy about cost, I thought I would use the break from intense discomfort to discuss options. "Jim," I said, "let's talk about what needs to be done. The next thing we need to decide is how to take a look at your kidneys. We can do what's called a helical CT scan. That's the most accurate. But it's also the most expensive. We can also do an old-fashioned test called an IVP. It's a lot less expensive, but..."

He had been resting with his eyes closed, but they snapped open, and he interrupted me. "You're talking cost?" he screamed. "I'm dying, and you're talking cost? How could you be so... inhuman? Do whatever is best. I just want this to be over!"

Medical care is easily viewed as an abstraction until it's yours. Then it's hardly abstract. I've never had a patient debate the cost of treating a kidney stone. When he or she discovers that something as big as an oversized grain of sand runs their lives, they experience an entire spectrum of emotions, but none of these emotions are very abstract. They are primal and immediate.

I would like to relate that Jim altered his philosophy after his emergency room experience. He did not. In fact, at his first post-hospital appointment, he brought in his hospital bill and waved it in the air. "You see, doctor," he said, "This is exactly what I mean about health care. Look at these charges!"

I was greatly tempted to remind him of what he had said in the emergency room, but that would have been inappropriate. When nature tortures human beings, they will say many things that they would otherwise not say—and that is exactly my point. Decisions about health care "purchases" are made by people in circumstances where they are at their most vulnerable and their least logical. Nothing can change that. Patients will behave like patients, not like consumers. Any claim otherwise is naive at best and hypocritical at worst.

Finally, let's consider our third reason. What if physicians were to behave like typical vendors? Let's consider vendors of the mythical widgets. They do everything they can possibly do to sell their widgets. In fact, if they don't do everything possible, then the system of supply and demand fails to work at its most efficient level.

For example, what if there are two sellers of widgets, and one of them suddenly develops qualms about how many widgets he is selling? He begins questioning consumers, "Look, do you really need this widget? Don't you already have enough widgets?" Obviously, some consumers will decide not to buy widgets or will go to the other vendor of widgets. That vendor is pushing widgets in every way imaginable and more consumers are buying his brand. Therefore, consumers conclude they must be better widgets.

The vendor of widgets without qualms thus finds himself in a position where he has a great enough competitive advantage that he no longer has to compete by lowering the cost of his widgets. Not surprisingly, he doesn't. The net result is that widgets do not reach their lowest possible market price and consumers pay more for widgets than if one of the vendors had not developed reservations about selling them to any and everyone.

A free market system assumes that vendors will sell their commodities using all means possible short of those that are deemed to be unethical—like claiming that not using a particular laundry soap carries a risk of dying. In fact, there are laws in place that prohibit vendors from *not* selling their commodities to everyone. If a widget vendor decided not to sell widgets to New York Yankee fans because he considered them minions of the devil, that vendor might well end up in court defending himself against charges of discrimination.

For physicians to meet their responsibilities as good and proper vendors, they have to promote the sale of their commodity—their services—aggressively and in every way possible. Every single illness, regardless of severity, should be touted as needing an appointment, every suspicious mole should be removed, and every cancer, regardless of how rare, should be aggressively pursued. Radiologists should promote new ways of visualizing body parts and develop the entrepreneurial skills to create imaging centers that compete with hospitals, and in general all new technologies should be widely promoted—not just to the medical community—as being vastly superior to older technology, creating entirely new markets. The possibilities are virtually limitless, and ...

What? You say that's wrong? You say that's the problem? You say that if doctors push their commodity, if they hawk their services, then things get out of hand?

As a believer in free market philosophy, I must remind you that it is not the responsibility of vendors to limit the sale of their services. It is the responsibility of the consumer. Weren't you paying attention when we went over what happened when the seller of widgets developed qualms about how many widgets he sold? Supply and demand lost its efficiency.

No, as a consumer it is *your* responsibility to know what is and is not needed for your care. You must *educate* yourself in your illness. You must become a *local expert in your own disease.* You see, supply and demand implies competition not only between vendors; it also implies competition between vendors and consumers. It is the job of vendors to move the economy along at a nice brisk pace by selling as much as possible at the highest price. It is the job of consumers to buy as much as possible at the lowest price, but not so much as to create problems. That's how it works.

What? You say that information about illnesses is confusing? So what? So is information about high-definition television, but you don't see anyone asking Best Buy or Circuit City to limit their sales or suggesting that maybe everyone should wait until costs go down to buy a new television. What would happen to the electronics industry if such a thing happened?

You're still whining? Televisions are one thing you say, your life or death are something altogether different? If your doctor is an aggressive vendor, a *vendor of services like any other vendor,* then who can you trust to advise you about what is and is not appropriate care? Good question. All the more reason to really study the medical literature. If you're really sick, maybe you should spend 10 or 12 years going to medical school so that you're truly up to snuff about what should and shouldn't be done.

By the way, we're working on deregulating commercial airline pilots. We think that one reason the airlines are having so much difficulty making ends meet is that they have to pay way too much to pilots. Aren't they just glorified truck drivers? Who says that a pilot needs to be certified to fly an airplane? Let the market decide. The airlines with the best pilots and the lowest prices will prevail. The ones that cut prices too far will crash a few planes, and people will stop using them. The market knows best.

Sounds a bit silly, doesn't it? But that's exactly what an unfettered free market has asked of physicians—to compete as typical supply-side vendors. It was quickly apparent that such behavior had insane consequences, and physicians were accused of being "the only merchants who controlled their own market." Economists and politicians sprained fingers in frantic efforts to point at doctors as the cause of rising health care costs. What did they expect? If doctors are no more than vendors of services, how are they supposed to behave? How can a free market ask doctors to compete with other doctors so the price of their services will drop and still ask doctors to limit the "sale" of those services?

Does the free market ask used car salesmen to sit every potential customer down and say something like, "Now look. I know you want to buy this 1997 Toyota station wagon, but we need to think about it. You're pretty young and you have three kids. You sure you can afford it? Do you even need a different car than the one you're driving?" Of course not! How can it ask doctors to do so?

Which brings me back to my third reason for health care's failure: Physicians *cannot* be vendors; they must be professionals.

Medical Metamorphosis

The business of health care is not business.

The secret of success in an institution...is to blend the old with the new, the past with the present in due proportion, and it is not difficult if we follow Emerson's counsel: "We cannot overstate," he says, "our debt to the past, but the moment has supreme claim; the sole terms on which the past can become ours are subordination to the present."

-Sir William Osler

CHAPTER TWO

HEALTH CARE IS NOT LAUNDRY SOAP

An office visit

I did not learn about health care by studying it as an abstraction. I was submerged in its intimate details on a daily basis. Most of you will learn about the system in the same way. You may end up in an emergency room following an auto accident, or be devastated by child's illness, or find yourselves overwhelmed by the Gordian knot of a parent's hospital bill. Health care's educational experiences come in an infinite variety.

Let's join the participants in one of those educational experiences. It's a reasonable way of beginning to look at how human nature impacts health care. It's a way of exploring reason #1 for health care's failure by immersing ourselves into the real world. We will join an encounter between a patient and his family doctor, an office visit.

It might be more exciting to consider the case of one of my own patients, a golf professional. One day while playing a round of golf with his son (who would go on to become the youngest golfer to qualify for the U.S. Open), he hit a solid 2 iron. He then collapsed, but not out of shock at having hit a solid 2 iron. As it turned out, he was rupturing his aortic root. The first part of the body's largest blood vessel was about to blow up. The efforts to save his life are heroic medicine that makes for spellbinding drama. It's the sort of medicine that television shows absolutely love. But it's far too easy to get lost in the drama and excitement of medical heroism.

So we will begin with a plain old office visit. The chances are much greater that you've experienced an office visit than that you've had a major blood vessel blow up. For that matter, they're probably greater than the odds of you having ever hit a solid 2 iron.

The story's characters are not unusual. The patients are a meld of a few of the patients I have cared for over the years. The doctor is a meld of a few of the doctors I have known with a bit of myself thrown in. We will use the story and some others like it as a template to discuss the nitty-gritty of health care and illustrate the validity of my three reasons for health care's failure.

As I tell the story, I'm going to ask you to change your point of view by stepping into someone else's shoes. For me, without a concerted effort to see the world from another's perspective, chances are I'm not going to truly understand what that person is saying. In real life, stepping into someone else's shoes is very difficult. Hopefully, in the less real world of a story, the effort will be easier. How much you learn will depend upon the truth of what I have to say, the skill with which I say it, and your willingness to consider another's point of view— particularly if it conflicts with your own.

In the first case, the shoes I would like you to occupy are those of Dan, a patient whose wife made the appointment for him. They are tan, size eleven, wing-tipped oxfords. Their owner is a forty-one year old male, who runs his own business and has two children. As he enters his doctor's office, you step into his shoes.

You are Dan.

Going to the doctor always makes you nervous, but this time you are honestly worried. You have—well, you're not sure what you have. That's why you're going to the doctor. However, as you enter his office, your smile says, "I really don't have to be here. I'm A-okay. In fact, the only reason I came at all is because my wife made me."

You look for a place to sit, unconsciously trying to find someone wearing a cast or bandage. Injuries are not contagious. On the other hand, people coughing or children with noses that look like weapons of mass destruction might possibly give you an illness which is AWFUL, CAUSING FEVERS OF 108, LARGE OPEN, FESTERING WOUNDS, AND MAYBE EVEN SOMETHING LIKE WHAT CAME OUT OF THAT GUY'S CHEST IN *ALIEN* ... AND...AND..."

Your racing imagination screeches to a stop as you notice an alert but quite desiccated man with a mean looking cane. He is staring at you, and you are sure he heard your internal voice of doom as it ran amuck. Your manhood is embarrassed by your fear. This guy has probably had more illness and pain than a character in a Stephen King novel. You nod to him, but he does not nod back. You're sure he's sizing you up, and you know you've been judged a wimp.

As you sign in, you try to turn down your emotions. You're relatively young, relatively healthy, and usually relatively in control of those emotions. Not today. You ask the receptionist how far behind the doctor is (not *if* he is behind), observe the stupid way she shrugs her shoulders in response, and decide to sit next to a teenager who looks completely healthy and normal except for various rings, pins, and other unidentifiable things sticking out of his person and the tattoo on his arm written in some unfamiliar script.

Once seated, you begin to feel less agitated. You even laugh at how this experience strips you of your normal sense of self confidence. You are uncomfortable with having become self absorbed. You puzzle over its cause. The pierced and painted adolescent elbows you and violates your cocoon of contemplation. "What's wrong with you?"

Alarms shriek. You are stunned by the primitive nature of his question. You stammer a bit, but it becomes apparent that his question was rhetorical. Before you can answer, he says, "I got VD. Got it from some slut I thought was clean. What a bummer."

You close your eyes, relieved and slightly angry at the same time. When you open them, your newly found confidant is slouched in the chair with his eyes closed. He is trying to go to sleep. You begin to wonder exactly how far behind the doctor really is.

Having learned that watching the clock only makes time pass more slowly, you do not look at your watch. You have taken the afternoon off, knowing that the appointment might take a long time. Also, should the appointment prove to be the first step in some tragic medical melodrama, you will not have to return to work.

You try to interest yourself in magazines, but they are tattered, old, and mostly devoted to subjects outside your areas of interest. For a moment you wonder about the selection of reading material, six issues of gardening magazines and five issues of Mechanics Illustrated. It occurs to you that all the others have probably been stolen.

You're finally called to go to an exam room. The medical assistant who takes you back is quite young. She is engaging, but her enthusiasm helps little. Your heart rate climbs. Your legs feel weak, as though you are walking uphill. She takes your blood pressure and notes that it is higher than usual. You nod with such resignation on your face that she reaches over and touches your shoulder in reassurance.

"It's not that high, really. It's just up a bit."

Nothing anyone says, other than your doctor, can assuage your fears. You smile a smile as bright as a toothpaste ad just to get her to finish her job so your doctor will come into the room and either confirm your worries or extinguish them. She leaves.

Now the wait becomes interminable. The tools of medicine—scopes, anatomic pictures, and an exam table covered with paper—surround you. You are distracted for a moment wondering if the paper is made specifically for the purpose of covering an exam table or if it could be ...god forbid—BUTCHER PAPER! Then you hear the door handle move a bit. The door opens, and in he comes—your doctor.

You have only been seeing this particular doctor a short time. You don't see him often, but you feel confident in him because you want to. You chose him because he was on the panel of your insurance plan, but a neighbor also gave you his name. You've heard others speak well of him. Right now, he is your doctor, and somehow, you are glad. He barely looks up from your chart and without calling you by name, he says, "Chest pain. You have chest pain?"

You're surprised by the lack of any social amenities and manage only a weak "Yes."

He continues studying your chart and says, "Okay. Tell me about it. When does it happen? What were you doing? What does it feel like? Do you sweat when it occurs? How about shortness of breath? Any of that? How long you had it? Is the pain severe?" He finally looks up, expectantly.

After the appointment, you wish you'd have said, "Anytime, anything, terrible, no, no, and three weeks, two days, eight hours and twelve minutes, and @$#% you!"

You instinctively recoil. The doctor's distant demeanor worsens your sense of vulnerability, and you protect yourself in the only way you know. You close off. You convert your fear into anger. You answer, "You know, I really don't think it's important, Doc. My family is worried, but hell, I've got life insurance. Why should they worry?" You laugh your best sarcastic laugh and spend the rest of the appointment shrugging off the doctor's questions.

It seems that the less you make of your complaints, the more he does. But your fear has now passed. You feel progressively more foolish as well as angry. You certainly don't want to take up the "good doctor's" time. He gives you the phone number of a cardiologist, saying something about having a stress test. You nod, but know you can get one done at your athletic club much cheaper than the co-pay charged at a specialist's office.

The receptionist calls you aside on your way out of the office. She says they do not have a copy of your current insurance card. You fumble through your wallet, but cannot find one. She apologizes, but says that without a current card, you will have to pay cash for the visit. You begin to question the policy, but there are now others in line waiting to check out. You pay for the visit with a credit card. The signature on the receipt is larger and a bit more slanted than usual. It resembles your signature on your son's last report card, the one with a series of "incompletes" and "Ds."

You are tempted to slam the door on the way out of the office, but the door is designed so that cannot be done. Some minutes later, you discover your fear is completely gone. All you feel is anger. It is many weeks before you realize you are the same age your father was when he died of a heart attack.

The pain continues to reoccur, but if the doctor is not worried, neither are you.

Having been in Dan's shoes for his appointment, I think you would agree that not even the American Medical Association's best advertising agency could turn your experience into a medical success. You have had symptoms that may be significant. You've tried to ignore them but doing so has made you anxious.

You finally allow your wife to make you a doctor's appointment. Though you don't really know the doctor very well, you are going to engage in an act of trust because you *want* to trust him. Your trust is violated. Under the best of circumstances, you would have had difficulty discussing your anxiety. The abrupt, distant attitude of the doctor made it impossible for you to even honestly discuss your chest pain.

You left the appointment feeling angry and discounted at many levels.

This is a failure of the medical system, but its cause is not terribly complicated. Basically the doctor in this instance is an uncaring jerk. Right?

Let's see. Let's have you step into *his* shoes. We will call him Doctor Jones. He is a family practitioner. He is fifty years old and has been in practice for eighteen years. He is married and has two children.

You are Doctor Jones.

Your stomach rumbles like a cheap cement mixer. Your eyes feel like they've been washed in Lysol. You ache in muscles you did not even know could ache. You desperately wish you could tolerate being on call the way that you did when you were an intern. Now, as a "well-established" and older doctor, loss of sleep wrecks you. Your prior night of call produced four phone calls from the hours of midnight to four A.M. Even when you get no phone calls while on call, you sleep the way you learned during your training. It's a sleep from which you can awaken with an immediate degree of alertness. It's also a sleep that leaves you exhausted and physically drained.

You try to pull your emotions back into some semblance of control. Across the exam room a long-term patient sits like a trapped animal, her eyes wide with fear as the reality of what you've just told her slowly soaks in. "I'm not sure I understand," she says. "Exactly what did the biopsy show?"

You answer softly, "The lymph node was positive for breast cancer. The damned thing is back again."

She visibly flinches. "It's over then, isn't it?"

You try to sound ultimately confident. "Absolutely not. There are some new drugs. Verna, look, this is something we'll fight. Don't think about picking out any eight thousand-dollar caskets quite yet."

She smiles. "You mean like my father's? How'd you remember that?" She is quiet for a moment. Then she says, "I was so damn mad how at much it cost."

There is another period of silence. Finally she shrugs, and looks at the floor. "I'm not much of a fighter."

You harrumph. "Horse manure."

She begins to rock in her chair. You reach over and touch her shoulder. "Verna, you need a few minutes. I'll be back."

You close the exam room door and hear Verna begin to cry. Ten years earlier, you would have gone back into the room. Now, knowing she needs time to cry, knowing you're behind schedule, you move to the door of another exam room. You feel anxiety climb into your throat knowing that even though you're running late you absolutely must return and spend more time with her. You hope your next patient has something simple like a sore throat.

The chart says his presenting complaint is chest pain. So much for simple problems. You open the door, and the wall goes up. It is detachment, a distancing from the patient. You know it's self-protective, but it's also against all your instincts. You hate it.

You enter the exam room while still reviewing the chart. Without even looking up or saying hello, you ask, "Chest pain? You have chest pain? Okay. Tell me about it. When does it happen? What were you doing? What does it feel like? Any sweating? Shortness of breath? How long you had it? Is the pain severe?" You finally look up.

"It's not that bad," the patient says and then goes silent. For a moment, he scowls, and then a measured smile crosses his face. "You know I really don't think it's important, Doc. My family is worried, but hell, I've got life insurance. Why should they worry?"

"Well, it was bad enough to bring you in here."

"My wife made the appointment." He almost sounds convincing.

"I see. How come?"

He shrugs.

"Look, your wife may have made the appointment, but ..."

"Doc," he interrupts, "I really don't think anything is going on. My old man died of a heart attack, but he smoked and was fifty pounds overweight. I don't smoke, and my weight's just fine. My wife is a big worrier."

"So you've had no chest pain?"

"None."

Seeing the patient's face recalls a flood of information. You rarely forget a patient's face—it is one of your quirky abilities. This patient had struck you as a pretty good guy. It's now very obvious he's angry, and you guess he's also frightened. You do your best to make up for your horrible entrance. Your emotional wall has collapsed because here is a human being, not just a name. All the strengths and weaknesses that brought you to medicine make it almost impossible for you to hide behind the wall when being face to face with a patient. You contemplate apologizing for being late, for being impersonal, but then you become apprehensive. The patient backpedals so fast that you have a hard time even doing an exam. You still don't know much about his chest pain.

"Let's start over," you suggest. "Tell me about your chest pain."

His response is clipped. "I haven't had any."

"You've had no chest pain?"

His smile tightens further. "Nothing important."

His anger is palpable. Were you not running late, were you not preoccupied by your last patient, were twenty phone calls not awaiting your attention, were you not exhausted from a night without sleep, you would find some way of deflecting that anger and establishing a connection with the patient. Then you might obtain a history that you trust. Today, you're incapable of the extra effort. You allow the patient to downplay the seriousness of his complaints. You allow yourself to participate in his denial and tell him if he does have any significant chest pain in the future, he should see you.

You try to bring appropriate closure by making a referral for a stress test. The patient races from the exam room, and you follow him almost as fast. You hurry back to see Verna, but she's already gone. You're concerned, but also relieved that you can try to catch up with your schedule.

Your day ends well after the dinner hour. You finish as much paperwork as you can. You still have ten phone calls to make, but you've answered the ones that appeared to be the most serious. You finally pick up Verna's chart and look up her phone number.

You call it, and she answers. It's apparent she's surprised by your call. As you review your suggestions for her care, you become increasingly uneasy. "Verna, I'll be honest. You sound like you're already dead."

"What?" Her voice is flat, devoid of even the animation of sorrow.

You search for words. "I said you sound like you're just waiting to die, like you're halfway there."

"Oh... really? I'm sorry. I ..."

"Is your husband there? I'd really like to see if he has any questions." Undefined concern starts to solidify into a specific worry. "I'd really like to talk to Bill, Verna. To see if he has any concerns of his own."

"I'm not sure if he's even home now, Doctor. He ..."

"Verna," you interrupt, "you been drinking? Lord knows I wouldn't be surprised. But I have to know, Verna. You've been sober for so long. Ever since your Dad died. We don't want to have to deal with booze and cancer at the same time. Damn, lady, that would be horrible."

There is silence for what seems like forever. Finally, she speaks. "What difference does it make, Doctor? Nothing makes any difference."

Your voice develops an edge. "Verna, I need to speak to your husband. Let me talk to Bill. If you don't let me talk to Bill, I'll call 911 and send out the police."

"What? Why would you..."

You interrupt again. "Verna, we've had a talk like this before, the last time you started to drink. And Verna, you tried to kill yourself that night. Let me talk to Bill, Verna."

You hear sobs, and you raise your voice again. "Verna, please, let me talk to Bill. It's going to be okay, but now I need to talk to Bill!"

You're finally able to talk to her husband. He's unaware that her cancer has reoccurred, but he confirms she has been drinking. He becomes increasingly angry. "Hell, she passed out. She's so damned drunk, she passed out. I..."

"Bill, how do you know she passed out from alcohol? Has she taken anything else?"

"Oh..."

"Bill, I'm going to wait here on the phone. You go look. See if you can find anything. Where does she keep her medications? Go look there."

"Oh, hell, she didn't..."

"Dammit, go look!"

You hear background sounds and muttering you can't understand but know is profane. As you wait, fatigue overwhelms the small surge of adrenaline that had come with the phone conversation.

Verna's husband finally picks up the phone. "Doctor, I have to hang up. I found an empty bottle of Valium from ten years ago and what looks like a brand new bottle of Tylenol—completely empty."

The Valium doesn't bother you, but the Tylenol does. "Call 911, Bill. Call 911. I'll alert the emergency room."

You hang up, call the emergency room and fill in the ER doc on Verna's history. At the end of that conversation, you are drained. You sit for a bit, literally summoning the strength to stand up. You put a note in your pocket with your chest pain patient's name and phone number on it. You plan to call him from your home, hoping that the extra effort of a phone call will shake loose a more honest history about his pain.

When you get home, you fall asleep in a chair. Your wife knows how grouchy you can be when you're this tired. She lets you sleep. The next morning, after having slept in your clothes, your skin feels like used sandpaper. You quickly abandon them to the laundry as you jump into a long shower. The note with Dan's name and number and a notation "chest pain?" remains in your pocket. It will become illegible after going through the wash.

Perhaps the doctor is not quite as great a jerk as he first appeared to be. Having stepped into his shoes, you must admit that he cared about his patients. But even well-meaning, caring physicians may fail to meet their patients' expectations if they are asked to do more than is humanly possible.

You've been in the shoes of both halves of an office visit, seen both halves of individual doctor/patient relationships. The one between Dan and Doctor Jones was a disappointment for both. But life is full of disappointments. Needs go unmet because of circumstances. Fortunately, this episode appears to have been without serious consequences.

Granted, should Dan continue to experience similar encounters in the future, he is probably going to receive less than optimal health care. Rather than risking exposure to uncomfortable circumstances, he will ignore symptoms.

With regard to Doctor Jones, when Dan shares his negative feelings about the office visit with his friends, he will wound the doctor's reputation. To some degree, it will damage the profession of medicine in general. Personally, should Doctor Jones continue to experience such encounters day in and day out, an obvious toll will be taken on his enthusiasm and his desire to care for patients.

For both, the encounter was troublesome but not disastrous. Right?

Unfortunately, in medicine "troublesome" and "disastrous" share a close kinship. Consider two postscripts to the office visit. Let's first step into Dan's shoes.

Once again, you are Dan.

After your appointment with the doctor, you return to your usual routine. The chest pain becomes a symptom that you explain away as heartburn. You take one of the highly advertised over-the-counter medications for esophageal reflux, and while the pain does not go away, you convince yourself it's improved. You become better at hiding the times when it occurs so that it is no longer a point of discussion between you and your wife. She accepted your cursory description of the doctor's appointment—"he said it's nothing"—perhaps because it was exactly what she wanted to hear. You're rarely ill, and the thought that there might be something seriously wrong with you frightens her.

You are still angry and embarrassed by your doctor's visit. You have tried to put it out of your mind and have never entertained either seeking a second opinion or returning to Doctor Jones. Things stay the same for some weeks. Then, while playing volleyball with your kids on a hot summer day, the pain returns. This time it is impossible to ignore or hide. You wait for a bit, hoping it will go away, but the pain is crushing. You sweat profusely. You are also overwhelmed by a fear that something awful is about to happen. You finally tell your family of your discomfort, managing a weak smile, trying not to frighten the kids.

Your wife calls 911. The paramedics arrive quickly, and quicker still perform the ritual of trying to stabilize the physiology of a dying man. Your children will carry forever the image of you waving to them as you disappear into the ambulance, mouthing the words "See you in a bit." You will be unable to keep that promise.

Now, step back into Doctor Jones' shoes. You are Doctor Jones.

You see that Verna is your next patient, and you smile. She was fortunate to have not incurred any liver damage from her acetaminophen overdose. Her brief alcoholic relapse resolved once her acute depression from the cancer's reoccurrence was addressed. Even more rewarding was her response to chemotherapy—few side effects, and she now appears free of disease.

You enter the room, and she immediately stands up. In her hands she carries a large plate of cookies covered with plastic wrap and a bow. "Hi Doc. I... I mean, my family likes these and I thought ..."

"Are you kidding? Cookies! Oh my!"

She smiles and hands you the plate. "I owe you more than cookies, Doc. You pretty much saved my life."

You shrug. "We were lucky, kiddo. Either that or the good Lord made me call you. I'm just glad I did. You look great."

You spend the rest of the appointment reviewing Verna's overall medical care. She kisses you on the cheek as she leaves.

With one patient left in your day, you get a call from the emergency room. You wait for the ER doc to come to the phone while munching on one of Verna's cookies. When he finally talks to you, the ER doc tells you that one of your patients, a 41-year-old year old male, came in DOA from his home, an apparent acute coronary syndrome. The patient's name does not trigger any memories. Your medical assistant brings you his chart. His face pops into your mind after you read the first few words describing his last office visit.

As you pore over the chart trying to define what might have happened, you suddenly recall his appointment in relation to Verna's. You then remember your intention to call him and how the phone call to Verna drove that intention to the back of your mind, filed with a thousand other good intentions. You feel your face flush and a growing hollow in the middle of your chest as guilt and failure begin to take their toll.

Verna's cookies sit on your desk untouched until days later when your medical assistant finally throws them away.

The postscripts are the stuff of nightmares, both for patients and for doctors. Patients wonder if their doctor missed some terrible problem. Having read about incentive programs to hold down costs, they may even question their doctors' motives. Quite simply, they wonder if their doctors care. Once patients' trust in their doctors is gone, nightmares are free to run wild.

Every bump is probably cancer. Every chest pain is probably an impending heart attack.

For a doctor, Dan represents the ultimate failure—a premature, preventable, and tragic death. Every day, doctors try to understand the complexities of the human body as it is buffeted by the human condition. They attempt this task while being hindered by many factors—time constraints, pressure to make each treatment not only the most appropriate but also the most inexpensive, and overwhelming paperwork.

They are also hindered by their own humanity. Society frequently overlooks this factor. So do doctors. To be human is to be imperfect, but society will often not accept a physician's imperfection. Dan's postscript defines a failure that is one of the most destructive forces a physician can face.

Whenever I hear someone take solace by saying, "Well, I should not feel so bad. It's not a matter of life or death," I wonder if they have ever pondered what it must be like to work in a profession where it *is* a matter of life and death. Of course not every office visit is a life or death proposition. Most are very, very routine. But Mother Nature can raise the stakes of an encounter in an instant without warning. When the stakes are raised as high as they were in Dan's case, could they possibly go higher?

Our office appointment was a routine medical encounter, yet it eventually resulted in the loss of a father and husband and a devastating blow to a physician. This is the fabric of health care.

It is a mosaic that is ultimately complicated because its patterns are created by the complex interactions of people, often in very stressful circumstances, facing life's most intimate and powerful issues.

It ain't laundry soap.

Medical Metamorphosis

"Every patient is a doctor...after his cure."

--Irish proverb

CHAPTER THREE

PATIENTS ARE NOT TYPICAL CONSUMERS

True confessions—the power of parental anxiety

When all is said and done, nature judges a species' success on one criterion—did that species leave a next generation? The biologic forces involved in producing offspring and caring for them are the most powerful that nature can offer. We are hardwired to pay attention to suffering in a child or spouse sometimes to the exclusion of our own. When your family is in pain so are you. One of the most difficult circumstances I had to face as a physician was watching the paralyzing fear and helplessness experienced by parents who had seriously ill children. It was a painful demonstration of the vast difference between a patient, or in this case the parent of a patient, and a consumer of services.

As a parent, I learned that for some of us even the threat of illness can be distressing. My first child was born the same year I entered private practice. Prior to my daughter's birth, I thought that as a doctor I would somehow be immune to the anxieties of a first time parent. Amanda had been home for but a few days when I realized what a ridiculous assumption that was. A parent is taught to be a parent by their first child. Without benefit of tutoring from these tiny professors of childrearing, anyone claiming to know anything about childrearing is, quite honestly, a charlatan. It is a field of expertise that demands hands-on experience.

My first year in medical practice I was a charlatan. I recall one mother's emotional plea about her newborn: "My baby has colic, and doctor I have no idea what to do. She cried all night and I rocked her, used a heating pad, put her in that swing thing that's supposed to calm down colicky babies, and...and...and... Help me doctor!"

My daughter, Mandy, recently had colic too. I had been up all night with her. But, being a well-trained professional and having had years of schooling, I answered the parent with great professional aplomb. "There's a swing thingy?" I screeched. "What swing thingy? What does it do? Where do you buy it? Come on, tell me. WHAT SWINGY THING?"

My second year of practice, I felt much less guilty about answering parents' questions, at least those about children up to a year old. Mandy was a very gifted instructor.

She was about sixteen months old when Mandy took me through the part of the course concerning illness in my own child. While washing her hair, I discovered a bump on the back of her head. I mentioned it to my wife who said she had noticed it as well. "What is it?" she asked me, assuming that as a physician I would give her an intelligent answer.

I smiled wisely and said, "It's a bump."

Somehow that answer did not satisfy her. "What kind of a bump?"

I waved off her question as though it was an absurd example of a first-time mother's hysteria. "It's the kind of bump that will go away," I said. "For heaven's sake, don't be silly." Inside, the part of my brain responsible for anxious parental worrying shifted into second gear.

The bump did not go away, and three weeks later the worry section of my brain was in fifth gear and racing along like Jeff Gordon on a NASCAR Sunday. I had scheduled Amanda an appointment with a pediatric surgeon. I had also made the back of her head inordinately tender with my constant palpation, trying to decide if this mass (no longer a bump) was in the posterior triangle of the neck, the area where neonatal neural tumors are most commonly found. I had continued to reassure my wife, but she was buying none of it. By the day of the appointment with the surgeon, she was genuinely concerned. By the day of the appointment with the surgeon I, on the other hand, was ready to puke.

I sat waiting in the surgeon's office, trying to make small talk with his receptionist as Mandy played with blocks, the only entertainment available (our waiting room had much better stuff). Of course, the doctor was running late.

He finally arrived, and we exchanged greetings as I swept Mandy off her feet and thrust her, head first, towards him. She giggled. He took a surprised step backwards, looked at the nurse's note about Mandy's complaint, and in self-defense felt the back of her head. "It's a bump," he said.

I opened my mouth to speak, but nothing came out. It was not a bump. It was a mass. How could he call something as serious as my baby's affliction a "bump"?

He noted my stunned reaction to his cavalier comment and said, "Jeff, it's a small congenital cyst. They're very common on the head. I suppose we could take it out, but..."

"It's not on the head," I snapped. "It's in the posterior triangle of the neck. You know, in that area..."

He cut me off by pointing to the back of Mandy's head. "Jeff, that's the head. Her neck is lower. It's the thing that attaches her head to her body." He then smiled, knowing exactly what I had been thinking. "She's just fine. Stop worrying."

I felt a flood of immense relief. I also felt very foolish. It was suddenly quite obvious that Mandy's bump *was* just a simple cyst. I felt stupid, but it was not stupidity that had stripped me of a sense of perspective. It was the fact that being a worried father trumped being a trained physician.

If a trained physician becomes less than rational when worried about a family member, what are the chances that the vast majority of people will behave as "intelligent consumers in a similar circumstance?"

Illness, pain and suffering, of the body or spirit, in oneself or a loved one, are unlike any other human circumstance. People in this circumstance are called patients or parents of patients or sons and daughters of patients. They should not be called consumers.

Medical Metamorphosis

A doctor, like anyone else who has to deal with human beings, each of them unique, cannot be a scientist; he is either, like the surgeon, a craftsman, or, like the physician and the psychologist, an artist. This means that in order to be a good doctor a man must also have a good character, that is to say, whatever weaknesses and foibles he may have, he must love his fellow human beings in the concrete and desire their good before his own.

-- W.H. Auden

CHAPTER FOUR

PHYSICIANS CANNOT BE VENDORS; THEY MUST BE PROFESSIONALS

McMedicine and its limitations

When I entered private practice, the only contract I had was with my patients. I gave them undivided attention, medical expertise gained through my training, and the benefit of experiential knowledge. In return, they paid me what was even then a modest fee. The charge for an office call was $12. The three doctors in my office had a total of four employees.

Thirty years later, I had multiple contracts with multiple insurance companies. Four doctors employed 18 people. I cannot tell you what the price of an office call was because it varied according to the insurance contracts, and often what appeared on a bill had little relationship to what I was paid. I still offered patients medical expertise gained through training and ongoing education. My experiential knowledge was obviously much greater. But patients no longer had my undivided attention.

Over time, my attention had been forcibly split. Health care's evolution into an industry had demanded that I become proficient in not just the operation of a medical office, but also in the language of complicated insurance policies, the process of negotiating with professional negotiators, and a variety of different sets of regulations covering everything from how a billing clerk submitted a bill to which hospital I could use. My attention was also drained by pressure to see an ever-increasing number of patients in a shorter period of time while meeting cost containment goals established by the insurance companies.

I called the net result of these forces *McMedicine*—fast food medical care.

When I began practice, one of my goals was to give my patients the same medical care I would have wanted for a member of my family. I can honestly say that for thirty years I did everything possible to meet that goal. But three years ago, I realized it had become impossible to give my patients the care I felt they deserved.

Late one night after a very long day, I stood on the deck behind my study and had what might be called a paradigm shift. I realized that nothing I could possibly do would ever stop the long days from growing longer and coming one after another. No amount of scheduling changes, no new and improved time management scheme, and no manipulation of employee responsibilities were going to make things better. Either I acquiesced to practicing McMedicine or I had to leave practice.

I turned to the huge old cottonwood tree in the corner of my yard and said, "Enough." I believe the cottonwood understood. The next morning, I tendered my resignation from the practice that along with my partner I had created thirty years earlier.

I have been asked, sometimes quite cynically, what could possibly be so different between McMedicine and the way I felt medicine should be practiced. I could quote Sir William Osler and wax prosaic in answer to that question, but an example explains it best.

I use a term I called "asking the question." It refers to the situation where a patient has a hidden agenda. The patient may not be aware they have this agenda because it may involve a fear or secret. The agenda may be revealed with an "oh by the way..." tossed over a shoulder as a patient leaves an exam room. It may surface in a phone call after an appointment that opens with, "I forgot to tell you something." It may also never surface.

"Asking the question" refers to efforts to expose the hidden agenda. There are many times when "asking the question" is the last thing a doctor wants to do. Hidden agendas are rarely simple problems. Dan, the patient with chest pain, carried a life-threatening hidden agenda into his office appointment. Unfortunately, it remained hidden.

Early in my practice, I was faced with a patient who had a hidden agenda, and I doubt I will ever forget its details. A new patient, we will call her Sherry, presented with a complaint of back spasms. She was in her late twenties, said she worked as a waitress, and lived alone.

Her past medical history was unremarkable, but the symptoms that brought her to the office were confusing. The nature of the pain seemed to change every time I asked a question. It had started during her morning shower, but not in a typical fashion like bending over to pick up a dropped bar of soap. She denied ever having it before and had done nothing unusual the prior day. She said the pain was severe, but her facial expressions showed no evidence of the pain. She also moved without apparent restriction. I did a physical examination and found nothing.

At the time, I had been in practice for three or four years and was still seeing a fair number of new patients every day. Some of these new patients were either drug seeking or wanted work excuses. I asked Sherry about needing either a note for work or any pain meds.

"No, I don't need either," she said. She turned away from me and looked out the window.

"Well," I said, "I guess you had a rather unusual intercostal muscle spasm. I have to admit, it presented a little funny, but that's what it looks like." Even to me, my explanation sounded lame. "You think you might need any muscle relaxants if it comes back?"

She continued to look out the window. "No."

"Any other problems, Sherry? Anything else going on?"

"No."

The office appointment had by then run over by about ten minutes. Although I had not really done anything for her, I felt I'd explored what there was to explore. "Okay then," I said. "If it comes back or you have any other questions, let me know."

Still looking out the window, she said, "Okay."

I stood to leave the room. I distinctly remember feeling confused and uneasy but at the same time reluctant to "ask the question." She was so distant that I was not even sure what the question might be. I stood at the door for a few seconds and then asked, "Sherry, you having any sleep problems?"

She did not answer.

"Sherry? Any sleep problems?"

"Yes."

"You thinking about suicide?"

"I have a loaded gun on the seat of my truck. When I leave your office I'm going to use it."

It happened just like that. Boom, whack, boom. I asked the right question and she answered—"I'm gonna kill myself." I

75

had no doubt she was serious, and indeed on the front seat of her truck we eventually found a loaded .32 caliber pistol. I can't remember exactly what I said next, but I do remember being grateful that it was said in a relatively calm voice.

I took her back to our staff lounge, and I placed her there with my MA (medical assistant). When I told my MA what was going on, her eyes did everything but fall out of her head. "What do I do?" she asked with a quiver.

"Stay with her, and if she tries to leave, tackle her and scream like hell."

"Are you kidding? She's bigger'n me."

When I returned to the lounge after calling a psychiatrist friend of mine, my MA was talking nonstop about her son, the weather, her mother's diabetes, and the cost of gasoline. Surprisingly, Sherry was listening intently.

The psychiatrist had stayed on the line, and Sherry agreed to talk to him. My MA and I left the room, but she was posted guard outside a half-closed door. I brought back my other patients one at a time and tried to reestablish some semblance of order to the morning's schedule. Fortunately, I was not then seeing thirty patients a day. I try not to consider what might have happened had Sherry come into my office five years later when I *was* seeing that many, when I was being asked to practice McMedicine.

Sherry agreed to meet the psychiatrist at his office. She refused to go to the hospital. I knew we were working on very thin ice, and I really did not want to precipitate a police call. So, going to the psychiatrist's office looked like the best plan, but I had to get her there.

In retrospect, I can't believe what I did next, but life sometimes works out for the best. I called Sherry a cab. When the cabdriver arrived, I told him the basic details. "Look," I said, "she can't go anywhere but to that doctor's office. She'll kill herself if she doesn't."

He didn't even blink. "Gotcha. Don't worry, Doc. She'll get there."

The conversation was held in front of Sherry, and at its conclusion, we both looked at her. She actually smiled and nodded.

I gave the cabdriver all the cash I had. "I hope that's enough," I said. I can't even remember how much it was.

"It's fine," he said. I'm guessing he would have done it for nothing.

Through my office window, I watched them leave the building and walk to the cab. The cabbie was walking close at her side and talking, almost as fast as my MA had. They drove off, and I said a short prayer.

Thirty minutes later the cabdriver called my front desk and told them she was in the psychiatrist's office. She was eventually admitted to a psychiatric hospital where a diagnosis of manic depression was made. She had been going through a period of what is called short cycling, big highs and big lows, one following upon the heels of the other. There is a significant rate of suicide associated with this state.

It was about eight months later when I next saw Sherry. She was in the same exam room where I had first met her. I was surprised to see her because I'd been told she was leaving Denver to return to her home in Arkansas.

I think I was somewhat more ebullient than was required. "Sherry! How are you? You look terrific."

She nodded and stuck out her hand. In it was $60 in fives and tens. "Here. This isn't all I owe you, because there's a charge for today. But maybe it'll cover at least the cab fare."

I started to say something about sending money later if she was going to be left short of funds, but I fortunately had the good grace to understand how wrong that would have been. "Thank you. That cabbie was an okay guy, wasn't he?" I asked.

"He was a real nice man," she said. "A real nice man."

"I'm glad you're alive, Sherry." I'm sure my eyes were holding back tears when I said that.

"So am I." Upon making this admission, she made eye contact with me for the first time.

Not much more was said, and she left. I've no idea why she chose me that morning. I've no idea why I chose to "ask the question" quite the way I did. Asking about sleep dysfunction is a standard part of looking for signs of depression, but why I jumped from there to a rather blunt question about suicide remains a mystery. I've no idea how well Sherry did after returning to Arkansas. Statistics predict that she faced a lifelong uphill battle against her illness. For my part, I have remained grateful that I chose to "ask the question," and that at least on that day more than twenty-five years ago, I was a medical professional, a healer, and not a just a vendor of services.

There is a difference.

The squeeze

In 1988, I wrote an article for *Medical Economics* describing how it felt to try to practice McMedicine. Unfortunately, almost 20 years later, the article is still relevant. The article produced a surprising degree of fame—not for me, but for Old White. Over the years, I have had many inquiries as to his fate and even a few sympathy cards.

A Rat's Eye View of Private Practice

I smiled as I read the card signed by my staff: "Congratulations on 10 years in practice! How come you got older and we didn't?"

As I looked around my office at the memorabilia of a decade, I could no longer conjure up the insecurity I used to feel as a brand-new family doctor. Now, I was a grizzled veteran.

The large white rat was an experienced research animal. It surveyed the box without fear, searching for the task for which it would be rewarded.

The letter was from the largest health maintenance organization with which I participate. It read: "We have noted your extensive use of consultants. As a result, we have reduced your reimbursement level from 80% to 70%. While the plan is not suggesting that your use of consultants is excessive, a decrease in this activity would result in a higher reimbursement level.

The rat looked at the round button, then turned to the square one. After a pause, it pushed the round button. As an electrical charge ran through the floor of the box, its body shook with spasms.

The noon lecture was on avoiding malpractice suits. It was the defense lawyer's third bullet point that made my head snap up from my study of the hospital lasagna.

"It is vital that you obtain consultations." His voice was low and dramatic. "If the thought of getting a consult even crosses your mind, get one immediately."

When the rat could control its limbs well enough to walk, it moved to the square button. It had long ago learned about positive and negative reinforcement. It pressed the square button. It received another shock.

The note attached to the chart was brief: "We have reviewed the chart and determined that the length of stay for this condition is seven days. This patient should be discharged tomorrow." It was signed by the Medicare Review Coordinator.

Hunger forced the rat back to the round button. It pressed down. The shock knocked it on its side.

The phone call was from Mrs. Iverholtz's daughter. "You've got to do something! She should be in the hospital for at least another week. If you send her home, she'll die—and it will be your fault!"

The rat moved desperately to the square button. It pounced on it. The shock was immediate.

I had two more letters sitting on my desk. One was from a private insurance company denying payment on a claim submitted 4 months earlier. It said the denial was because electroshock therapy had to be pre-certified by their psychiatric review committee and that I had never done the pre-certification. The claim had actually been for fracture care for a broken arm. I have never administered electroshock therapy in my life.

The second letter was from Medicare concerning a different patient. It said I had provided an unnecessary service, violating the law. It cited possible criminal penalties should I be convicted of violating the law. It took me a moment to figure out that the letter had been generated because I had done a Pap smear on a 65 year old woman. Apparently the reviewer thought the need for the test stopped along with menses.

The rat sat immobile in a corner of the box. It stared distantly, its red eyes unblinking.

The research psychologist shook his head. "Even Old White couldn't handle that. It'll break the best of them."

--Reprinted with permission from *Medical Economics*, Nov. 7, 1988, p. 160. *Medical Economics* is a

In the past 15 years, there has been increasing pressure applied by quality assurance organizations to guarantee that physicians are up to date in their knowledge and procedural skills. Because of this, most specialty boards have followed the example of the Academy of Family Practice and required ongoing education and recertification. There is no doubt that such activities promote the general level of knowledge within the medical community, but it is the *performance* of that medical community, the manner in which doctors perform their day-to-day activities, that determines quality of care and the overall number of medical errors.

In 1988, Old White demonstrated what happens when even "the best of them" faces overtly conflicted goals and an ongoing series of lose/lose dilemmas. Circumstances have only worsened in the ensuing two decades. It is impossible to overestimate the negative impact these circumstances have had on quality of care.

Regardless of how talented or how well educated physicians are, they will not perform very well when spending every day dreading the next jolt of electricity. And a doctor who is worried about that next zap cannot deliver high quality care to his patients.

Medical Metamorphosis

"If you ain't got your health, you ain't got nothing."

--Anonymous.

CHAPTER FIVE

DENYING THE OBVIOUS

Primal issues

Pain is nature's way of telling an organism that something is wrong. The brain rarely ignores pain. "Yo Dumbo! Get your hand off the stove NOW! It's still hot" is a message that will focus your attention even faster than realizing that the twenty dollar bill looking up at you from the gutter is real. As far as your mind and body are concerned, the immediate threat to the integrity of the skin on your hand is infinitely more important than unexpected treasures, even if it was a hundred dollar bill in the gutter.

Being able to sense pain is even crucial to survival. Children who have a congenital absence of pain live substantially shortened life spans. Because it is such a primal sense, pain's traveling companions, death and suffering, are core issues of art and religion. They have preoccupied humanity's attention since the first recorded word. They are the cornerstones of humanity's perceptions about the universe. All else follows.

I know this is not exactly a revelation—"Hey honey, come read this. This guy says people don't like pain, that when you put your hand on a hot stove you take it off fast because your body thinks it's really important not to cook your fingers. How about that?"—but with regard to our subject, health care, it needs to be underscored. Why? Because it is a reality often ignored by health care experts—economists, social scientists, politicians and their brethren.

Tweak our example of the stove and the money in the gutter. Place wealth and pain in the same picture. Instead of a stove, make the source of heat a fire and allow a gentle breeze to blow a winning lottery ticket into the fire. The ticket's owner will reach for it, but the chances are overwhelming that he will yank back his hand, and the ticket will go up in flames. He has made a choice based upon the protective mechanism of pain that has profoundly influenced his financial status. Economics were trumped by pain, a primal force, by the human condition. You don't have to look very hard to find myriad stories written about this theme, but it is a theme the health care experts ignore.

Therein lies a primary reason why the first step towards curing health care's crisis is not taken—simple realities, like the one's we've explored, are ignored. Their power is denied.

Even smart people do it

It would seem that much of what we have been discussing is so obvious that it *must* be a part of all health care theory.

Unfortunately, as in the example of Jim, our politician who had the kidney stone, denying the obvious is all too easy, all too common.

Richard Lamm is a former governor of Colorado. He is thoughtful, articulate, well read, and has the courage to address issues that most politicians avoid. Health care has long been one of his concerns. He came to national attention some years ago when one of his opinions was interpreted as a suggestion that old people "have a duty to die and get out of the way." (He never really said that.)

Lamm has written a book entitled *The Brave New World of Health Care* (Speaker's Corner Books, Fulcrum Publishing, Golden, Colorado, Copyright 2003). It is well thought out and demonstrates an excellent understanding of many of health care's issues. However, it is also an example of how the primal nature of health care can be ignored and how treating health care as purely an abstraction can be enormously deceiving.

The opening line in Governor Lamm's book is, "My wife, Dottie, survived breast cancer. I am forever grateful to the skill and caring of an excellent group of doctors and a good hospital. There is no doubt in my mind that we owe her health and her life to the brilliance of the U.S. health care system."

He then ignores this episode for the remainder of the book and goes on with the theoretical opinions of a former governor. He characterizes his point of view as a "macro," as opposed to a "micro" perspective of an individual physician or patient. He emphasizes his belief that the system can no longer afford to cater exclusively to the individual patient, that the "good of the many" must supersede the "good of the one." He's not afraid to create his own sound bites, like "You can't fund a health care system a mother at a time," his message being that America can't afford to deliver health care the way we would all want it delivered to our mothers. He adds, "Painfully, but inevitably, we must put our individual needs within the context of other patients' needs and other social demands of a complex society." Governor Lamm even has the temerity to consider rationing health care. Finally, he discounts the doctor/patient relationship, characterizing it as being problematic because it is incapable of self-regulation.

But he fails to ask, much less answer, a crucial question— if someone had suggested that treatment be withheld from his wife, Dottie, because of cost or abstract societal concerns, what would have been his response? What would he have said if he had been told, "We think we can not only save your wife's life but also assure her of many disease-free years. However, the treatment falls outside the standard list considered acceptable and is too expensive."

I suggest that his response would have been like that of any other loving husband. He would have snarled and growled and fought to get his wife the best care possible. Rightly so, because she *did* get the best care and twenty years later is still living a productive life. That's why Governor Lamm begins his book with a tribute to American medicine. He did not begin it by writing "When Dottie got breast cancer, we had a hard time justifying the cost of her care because she's not as important as an abstract consideration of the allocation of society's resources." ARE YOU KIDDING ME? What he wrote was basically, "Man, am I glad that excellent group of doctors and that good hospital saved my wife's life so that twenty years later she is alive and still trying put up with me." (My words.) Whatever else he says about cost or societal priorities follows that admission, that tribute to medical care.

I respect Governor Lamm's personal history of intellectual courage and creative thought, but in this circumstance he is a hypocrite. His hypocrisy lies in abstractions that belie the intimacy and power of illness, suffering, and pain.

For example, in the abstract, as Governor Lamm points out, rationing of health care is one way to control costs. When renal dialysis was a new treatment modality, its use *had* to be rationed. There were more patients in renal failure than there were dialysis units. One of my patients, a protestant minister, sat on a board whose function was to decide who received dialysis and who did not.

One day he told me, "Doc, it was undoubtedly the worst experience of my life. How do you decide between a young father of three children and an older doctor, the only doctor in a town of eight hundred? The father worked in a shoe store, not what you'd call a job essential to the good of society. The doctor,

89

well, that town really depended upon him. But he was old and was going to die sooner than the father, and the father's kids depended upon him for *everything*. How do you decide? The whole board quit after a year. None of us could handle it. Nobody should have to make that kind of decision." My patient had experienced the difference between abstraction and reality and what happens when that difference is ignored.

The rationing of dialysis was a short-term experiment. It became unnecessary when the health care system massively expanded its dialysis capacity. Today, treatment of renal failure represents one of the more expensive treatments paid for by Medicare. However, dialysis is not one of the areas being attacked as an inappropriate treatment modality. True, the doctors who perform this service are constantly being squeezed, but doctors are the easiest part of health care to squeeze, poorly organized and not terribly comfortable defending themselves. None of the experts suggest dialysis be rationed. Why? Because of the experiences like those of my patient.

The tragedy for all of us is that the experts appear to have forgotten this episode. They appear to have learned nothing from this experiment in dealing with life and death as an abstraction.

You might consider axial CT scans to be an over utilized, expensive radiological study until the surgeon says he thinks one will help decide if an emergency appendectomy is needed for your eight-year-old daughter who is writhing in pain on the ER bed. Then they become a miracle of modern technology.

You may view statins as being inordinately expensive medications, felt by some to be used inappropriately, until you are told to take one because your cholesterol is elevated, your father died of a heart attack when he was forty-nine, his father died of one when he was fifty, and you are now forty-two-years old. Then, statins are the lifesaving product of hardworking scientists.

Hip replacements in the very aged are cited as a huge unnecessary source of Medicare expenditure, and your ninety-four-year old mother just fractured her hip. She happened to fracture it while riding her bicycle, and she's still sharp as a tack. Are you going to tell her she's too old to have it replaced and should go sit in a chair and wait to die?

When Governor Lamm described what I assume was his own family's most significant health care crisis and then ignored the power of that crisis for the rest of his book, he claimed immunity from individual reality. He placed himself in a state of denial. He simply ignored a circumstance that did not further the arguments he wanted to make.

Even if Governor Lamm would *not* have done everything in his power to get his wife treated, which I doubt, I can assure you most husbands would have. What happens if every patient denied a treatment protests that denial? What happens if their doctors protest as well? Can a medical system function if neither half of the doctor/patient relationship supports it?

Medical Metamorphosis

You don't have to look very far for an answer. When patients felt HMOs were sacrificing lives so that insurance companies could make increased profits, there was political hell to pay. Eventually, federal legislation was proposed to protect patients' rights. This all happened within a few short years of the rise of HMOs. Any significant change in the health care system must acknowledge the power and intimacy of the commodity with which health care deals or the implementation of that new system will be chaotic and doomed to failure.

The business of health care is not business because "if you ain't got your health, you ain't got nothing."

STEP TWO

**CREATE AN INFORMED CONSUMER, CONTROL
HEALTH CARE COSTS, FACILITATE SUPPLY AND
DEMAND, INSURE ALL AMERICANS, AND PROTECT
THE INDEPENDENCE OF MEDICAL PROFESSIONALS--
WITH *ONE* CHANGE--
DEMAND DIRECTED HEALTH CARE (DDHC)
and a
REQUISITE INSURANCE SOURCE (RIS)**

I am sure there are some of you who have dismissed my explanation of why supply and demand could never work for health care. There are probably even some of you who are hoarse from scoffing. I am hardly a world renowned economist offering powerful theory.

But I would ask all you doubters to do one more thing before turning your attention elsewhere. Look back to the last time a loved one was ill. At any point during their illness did you approach their treatment in the same way you approach the purchase of a new television or a piece of clothing.

Did you at any point ask the doctor, "How much will this cost?" and then do a price comparison at another hospital or with another doctor? Was price ever even on your mind or was your mind preoccupied with worries about your son or daughter, mother or father, or wife—whoever was ill?

I absolutely know that the vast majority of you, if honest, would answer those questions by admitting that cost was never much of a concern.

Therein lies the proof of my thesis.

Many of you will also assume that I will next argue for socialized medicine, a single party payer, or some variant of the national health systems of Great Britain and Canada.

You're wrong. These systems are struggling because they are often inefficient and unproductive. They are laboring to find ways to inject *competition* into the supply side, into hospitals and groups of health care providers.

Neither an unfettered free market nor a closed national health care system work very well. What is needed is a system that fosters competition but acknowledges the unique nature of health care. Such a system is possible by applying the tenets of Demand Directed Health Care through the creation of a Requisite Insurance Source, step number two in the cure of America's health care crisis.

The most meaningful lessons will be those conveyed in the context of ...relationships in which trust has been earned by demonstrated scientific rigor, by devotion to patients and colleagues, and by a track record of adherence to a higher law that excludes the self-serving motivations so prevalent in our culture and, in many cases, our profession today.

--- J. D. Mellinger, *Surgical Endoscopy,* (2003) 17: 363

CHAPTER SIX

LIFE SAVING SURGERY

America's disgrace

When the Emancipation Proclamation was signed, America had already invested tens of thousands of lives in its cause. Yet right up until that moment, there were those who cited complicated political and economic reasons why it should not be signed. History has rightfully judged some of those who opposed its signing as being self-serving and duplicitous. An irrevocable down payment in blood had been made, but they were still too invested in their own self-interests to commit to slavery's abolition. Others simply lacked the courage to commit to an act of such magnitude even though in their hearts they knew it to be morally correct.

95

Medical Metamorphosis

America's health care crisis and the loss of life suffered because of neglect obviously pale in comparison to slavery, the Civil War, and the potential dissolution of the Union. It is no less reprehensible. And let there be no doubt, the only reason this country does not have health coverage for all its citizens are money and fear. Any attempt to deny this fact is pure fabrication.

To claim that health insurance is a privilege and not a right is shaky rhetoric. The Declaration of Independence declares "life, liberty, and the pursuit of happiness" to be inalienable rights. None of those three rights are achievable if a man or woman must live under the yoke of pain and suffering. And no amount of erudite rhetoric can obscure the fact that death makes *all* of a person's rights rather irrelevant.

It is incomprehensible that there is not a universal health insurance system in the world's richest country. How can we tacitly accept the deaths of the 500,000 Americans estimated to have died since 9/11 because of our health care system, 80,000 of whom died because they simply had no health insurance? The most common answers discuss the complexity of the health care system, or the possibility that most solutions will engender opposition from one or another of the "big players," and lectures on the historical background of our present debacle. Those answers are no more acceptable than were the excuses offered for failure at all levels of government in their responses to Hurricane Katrina. Five-hundred-thousand lost lives represent many, many Katrinas. America is being buffeted by a class five hurricane of neglect and hypocrisy, greed and self-interest.

It's no longer acceptable for economists, social scientists, politicians, and academicians to shake their heads and say, "You really don't understand the complexities and financial implications of dramatic change. It's not that simple. You don't understand." Those who died because they lacked health care were just as hopeless and abandoned as were the victims of the deafening tragedies of 9/11 and hurricane Katrina, just not as visible. No death is so anonymous that it does not matter.

John Donne knew that when he wrote, "No man is an island, entire of itself...any man's death diminishes me, because I am involved in mankind; and therefore never send to know for whom the bell tolls; it tolls for thee."

If you are unfortunate enough to be without health insurance, the health care crisis carries a financial risk that often appears greater than the risk of poor care or no care at all. Medical bills are cited as the second-greatest cause of personal bankruptcies, and following recent changes in bankruptcy laws, debtors' prisons may no longer be a thing of Dickensonian England. It has already taken legislation to stop hospitals from foreclosing on the personal homes of uninsured Americans who had the audacity to become sick enough to require hospitalization. If you don't have health insurance, you're praying to stay healthy because your hospital bill may be *ten times* that of an insured patient with exactly your same problem and treatment. Those who can least afford it are charged the most.

Some health care experts are also trying to shift blame onto patients' shoulders by suggesting that those without insurance would be much better off if they just lived a healthier lifestyle. It's probably true that if you don't have insurance you're not knocking yourself out doing what you should to stay healthy. As a matter of fact, if you don't have health insurance, you're probably beating your brains out just trying to keep a roof over your head.

Balanced meals? Twenty minutes of brisk walking per day? How about some gentle stretching, a massage, and a session with a personal Zen trainer? YOU BETCHA! Instead, how about one meal a day grabbed at a fast-food joint, a race through traffic to a second job in a 1985 clunker that smokes like it's on fire and burns a quart of oil every two days, a gallon of coffee to stay awake, a stagger up the sidewalk, a collapse into bed, and five hours of sleep—which barely qualifies as sleep because of the coffee. Asking people whose lives is a variant of this chaotic battle to pay attention to a healthy life style is stupid. It's not stupid because it wouldn't help people stay healthy. It's stupid because they won't or can't do it.

The same is true with regard to affording health care. If you are the person hitting the convenience store Mondays and Thursdays to buy an overpriced quart of oil to keep your car running, your first goal is probably to get a new car. Problem is, you're spending so much on oil you can't afford a new car.

Health care does not become a priority until you're emitting smoke from your own exhaust and burning oil as badly as is your car. By then, you're probably sicker than your car. Unless something is done at a societal level, there will always be many, many people who can't afford health care. Preaching that they should find some way of doing so just doesn't cut the mustard.

It's not a particularly sound financial proposition either. When people finally become really sick, they *will* seek care somewhere, usually in an emergency room. Even though America lacks universal health coverage, it is not yet a society able to step over the bodies of heart attack victims whose employers recently dropped their medical benefits. "Good Heavens, Buffy, that fellow clutching his chest is a dreadful shade of purple. How inconsiderate. He should have the social grace to be home turning purple and dying instead of here—in the parking lot—with that quart of oil clutched in his hand."

That won't happen. Someone will call 911. An ambulance will arrive, the quart of oil will be pried from the hand of our uninsured patient, and he will be taken to the emergency room. There, marvelous and expensive technology will stabilize a condition that could have been avoided had this particular patient stayed on his blood pressure medication. Using emergency rooms to treat simple problems or having to pay for avoidable and catastrophic complications of untreated common diseases are just the tip of a financial iceberg that could sink a thousand Titanics.

But any suggestion that there should be some form of

health insurance for all Americans is immediately countered with massive amounts of hand-wringing, head shaking, and muttering about the incalculable problems. But just think about it. Universal health coverage is about protecting American lives, both in quantity and quality. Americans *do* die because they don't have access to health care, and they most assuredly suffer if they lack it.

Until someone honestly admits this truth, ends this intolerable hypocrisy, and approaches universal coverage *with the same resolve and courage that was required to invade Iraq*, debate will continue and so will the suffering and loss of life. Nothing short of such an act of courage, commitment, and resolve will end the needless loss of life, the tragic pain and suffering.

Are the lives lost because of a lack of universal health insurance worth less than those lost on 9/11? Since that date in 2001, there have been a half million lives lost because of America's health care system. This is more than one hundred times the number of lives lost in terrorist attacks during the same period. A half million lives is a small American city—gone, buried, its citizens casualties in the war on disease and suffering whose losses were barely noted.

Regardless of how universal coverage is initiated, administrated, and paid for, there will be difficult compromises, confusing changes, and a painful challenge of long-standing beliefs. But I have yet to hear an argument that comes close to convincing me that these problems would be worse than those associated with America's present medical industry.

If you are an American without health insurance, which is a greater risk—dying in a terrorist attack or not having access to medical care? If the health of American children is not a priority, what is? If health care is a privilege and not a right, what *is* a right? And in the end, how can America tolerate the ethical hypocrisy of this circumstance and still claim to speak to the rest of the world with a voice of moral authority and compassionate concern?

Willie's lesson

Some years ago I had a patient we'll call Willie. Willie was a sixty-two-year old man who was a tough, stubborn five feet two inches tall. He had thick, silver hair combed straight back and a pencil-thin mustache. He also had a left leg with almost no pulses. Arteriosclerosis and diabetes had severely narrowed the vessels carrying blood to the leg, and it was dying.

The first signs of necrosis—death of tissue because of a lack of blood supply—occurred in Willie's fourth and fifth toes. They basically turned black. Willie was not terribly fond of seeing doctors. He was even less fond of seeing doctors who suggested that he have two of his toes amputated. I sent him to three different surgeons. They all said the same thing—the toes had to come off and something had to be done to increase blood flow to his leg.

Willie responded to these suggestions by saying to each of the surgeons, "Okay. I'll think about it. Oh by the way, just curious, what happens to the toes if I don't do anything?"

The surgeons replied, "They will fall off."

Willie nodded thoughtfully. "I see. Just asking."

I know this to be true because it was an exchange also played out between Willie and myself, four times, identically.

One day Willie came in for an appointment. It was apparent something had happened because Willie had made the appointment himself, without any arm-twisting by his wife. "Hi Doc," he said. He began to untie his shoe. "Guess what? They fell off. Just like everybody said, they fell off."

He whipped off his sock and sure enough, the black fourth and fifth toes, or most of them, were gone. At the tip of their stubs was a hardened area of even blacker, dense tissue, the residual of a process called auto-amputation. "How about that, Doc? It didn't cost me or Medicare a dime."

Willie gloated for a few weeks about the economic and medicinal virtues of benign neglect. When I broached the subject of the rest of his leg, he dismissed my concerns. "I'll think about it," he would say as he limped out of the exam room.

Then one day he came in for an appointment with his face contorted in pain. "Doc, I think things might be worse," he said through clenched teeth. "My foot, my leg, holy moly, they're killing me."

I helped him remove his sock and winced when I saw

severe mottling of his foot. "Willie, things are worse all right," I said. "Now your whole foot is dying. I'm not sure you want to wait for your whole foot to fall off."

Willie finally agreed to have an angiogram of his leg—a dye study of its blood vessels. It demonstrated two or three very narrow areas of stenosis, but, as the radiologist noted, "all the blood vessels pretty much looked ratty."

After the angiogram, Willie was upbeat. "Well, that was not fun, Doc. But I'm glad I did it. So what medicine are we going to use to get this thing cleared up?"

I shook my head. Willie's ability to see the world only as he thought it should be was a consistent amazement. "Willie, look. Let's go over everything again. You're already on every medicine we have available to open up blood vessels, to make your platelets slippery, and to control your diabetes. They're all doing about as well as they can. But your blood vessels are really shot. We did that study to see if maybe surgery might help."

"Surgery?" The tone of Willie's voice made it sound as though surgery had never been mentioned even though the day before the angiogram we had discussed the possibility of a bypass graft. "What sort of surgery?"

"Remember?" I tried not to sound impatient. "We talked about a graft, a tube of Dacron to replace the narrowest areas of your big blood vessel."

"Oh right. I'll think about it."

I decided that I had to raise the ante; otherwise Willie

might bluff himself into a disaster. "Willie, there's one other thing we might want to consider. I've got a pretty good idea what you're going to say, but we need to talk about this anyway. It may actually be the best alternative. We may have to consider an amputation."

He shrugged. "Toes? Part of the foot?"

"The leg, Willie."

Sometimes Willie pretended that he'd missed part of our discussions because of a very selective hearing deficit. He made no such pretense when I mentioned amputating his leg. "Doc, that's not even a possibility! Holy moly, are you crazy?"

There then followed a period of a few weeks during which Willie rapidly went downhill. His leg became progressively more painful. He developed a number of skin breakdowns. In addition, his diabetes became increasingly more difficult to control. The stress of his dying leg was taxing his entire physiology.

He decided that a graft would cure everything. He was steadfast in this belief because he was completely opposed to any whisper of amputation. Unfortunately, it had become my opinion, and the opinion of two vascular surgeons, that Willie's disease was too diffuse for a graft to be effective. But Willie's mind was made up. He went to a limb preservation clinic for a third opinion.

A limb preservation clinic is exactly that, a group of specialists and therapists whose goal is to do anything and everything to see that patients keep their limbs. While this appears to be the noblest of goals, there are circumstances where efforts to do so compromise the rest of the patient. I felt Willie was one of these circumstances. The suggested treatment for saving his leg was extensive surgery, not one but two grafts and another piece of surgery called a sympathectomy, surgery to open the smaller blood vessels by eliminating the nervous input that constricts them. I was not just concerned that these surgeries stood a relatively small chance of helping Willie. I was also concerned he might not survive them.

I explained my concerns to Willie using every metaphor, example, and statistic I could think of. His response was "I'll think about it." Two days later he called me and said "Surgery's the best way to go, Doc. What do I do now?"

Once I was convinced that Willie was convinced, my job changed. I did everything possible to get him tuned up and ready for surgery. He had a pulmonary and diabetic consult. Because of the number of specialists involved in his care, I made a particularly aggressive effort to be sure they were all on the same page.

Willie first had the sympathectomy. Outside of a complaint that his legs felt like they were on fire because of the superficial flushing of blood vessels, everything went fine. Then came the two grafts, done at one sitting. The surgery itself took eight hours. Twelve hours after it had been completed, the surgeons had to go back in because one of the grafts had clotted off. Twelve hours after that, the leg itself had to be opened back up because of bleeding—the place where the graft had been attached had broken down.

Thirty-six hours after the surgery, Willie had still not regained consciousness. The outcome looked grim. The vascular surgeons were considering a complete revision of his bypass grafts. Willie's family immediately vetoed that idea without me even having to lobby against it.

At the forty-hour mark post-op, I visited Willie's room in the ICU. His wife and two daughters were at the bedside. The daughters looked exhausted. His wife simply looked devastated. She was beyond exhaustion.

"Oh, Doctor Waggoner," she said. "I never thought this would happen. Willie was so sure this would make everything better. What happened?"

I suppose she might have asked a more loaded question, but I can't think of what it might have been. But in this circumstance, the simplest answer also offered Willie the greatest chance of surviving. "Well," I said, "Willie wanted to take a chance on keeping his leg and still improving his quality of life. Unfortunately, that effort failed, so now we need to get on with it."

She wrinkled her nose. "Get on with what?"

"His life. His life and what we have to do to save it. We need to amputate that leg, and we need to amputate it now."

She continued to wrinkle her nose, turned her head

towards the ceiling, and closed her eyes. "Oh no, no, no. He'd kill me if I let you cut off his leg."

I had an easy response to that concern. "You'll kill *him* if we don't."

We amputated the leg. Three hours later, all his vitals stabilized. He was awake and alert seven hours later, and having a meal two hours after that. Willie's leg had been a terrible burden on his entire metabolism, a huge load on his cardiovascular system, and a depressant in general because of the chronic pain it had caused. It is not an exaggeration to say that after the amputation, Willie was a new man. When I left my practice, he was ten years out from his amputation and doing better than I could have ever hoped.

Willie's case serves as an appropriate metaphor for American health care. Much of the industry is like Willie's leg—dead, nonproductive, and lethal to the system as a whole. There have been a number of unsuccessful attempts by the experts to save those dead parts. Willie's diabetes, pain, and overall health steadily declined in spite of sophisticated medical treatment and, finally, surgery. Health care's cost control, quality, and compassion have deteriorated in spite of HMOs, restrictive legislation, and the creation of a new industry whose only function is to critique and criticize medical treatment. To save Willie's life, his leg was amputated. My opinion is that the same is true for health care. Until its dead parts are amputated, there is little hope for improvement.

Willie refused to accept losing his leg as being necessary for his continued survival. In fact, he never did agree to the amputation. His family made that decision while he was unconscious. Unfortunately, this same self-destructive resistance characterizes the health care system. Well-entrenched and diverse forces—insurance companies, private hospitals, organized medicine, trial lawyers, politicians (both conservative and liberal, Republican and Democrat), the best-paid medical specialists, and medical academicians—are intensely protective of their agendas and fiefdoms. This self-serving stasis is the source of hand-wringing, finger-pointing, and scapegoating.

These "players" scoff at proposals for dramatic changes in health care. They declare categorically that anything other than "incremental change" is doomed to failure. The first evidence they typically cite in this regard is President Clinton's (usually they blame his wife) failed efforts to reform health care in 1993. Discussion stops at that point. Why? Have circumstances not changed since 1993? Is it not possible that the efforts in 1993 were a failure of style and not substance? In the interim twelve years has incremental change improved anything?

Those who now disdain any change other than incremental are the same forces that defeated reform in 1993—insurance companies, hospital corporations, pharmaceutical companies, and organized medicine. These are the forces that stand to gain the most from circumstances remaining as they are. They dismiss plans for an overhaul of the health care system knowing they are the very powers that will smother them. Self-interest gives birth to self-fulfilling prophecy.

Medical Metamorphosis

What makes it so plausible to assume that hypocrisy is the vice of vices is that integrity can indeed exist under the cover of all other vices except this one. Only crime and the criminal, it is true, confront us with the perplexity of radical evil; but only the hypocrite is really rotten to the core.

--Hannah Arendt

CHAPTER SEVEN

THE SUIT DON'T FIT!

The clothing salesmen

Let's return to our parable of the ill-fitting suit. Before we talk about what a *well*-fitted piece of clothing looks like, let's take a look at what America has been asked to wear for the past two decades. Let's also look at those who shoved America out the door, tripping on pants that are too long while holding up one sleeve and tucking another against its chest.

In our parable, the clothing salesman accomplished his goal—selling a suit—by forcing the customer to perform gymnastics to make the suit fit. That is exactly what the "players" have done with regard to health care. They have asked Americans to carry out emotional, economic, and intellectual gyrations, not for their own benefit, but to meet the needs of the players.

111

Who are these players—these clothes salesmen?

The "players"

I have a friend who has summarized his attitude about the health care system quite succinctly, "I am a capitalist, but capitalism does not do health care very well." There is ample evidence that he's right, but why does America continue to force the issue? After thirty years, why does the experiment continue?

There are four powerful players in the health care "game"—for-profit insurance companies, for-profit hospital corporations, pharmaceutical companies, and organized medicine. If one investigates the financial performance of these industries, the answer to that question is readily apparent.

The "Fortune 500" is a ranking of America's largest 500 companies. It is released annually by *Fortune* magazine. The study also evaluates America's top 50 industries within the list. Consider the evaluation of for-profit insurance. In 2006, when ranked for total return to shareholders, this industry ranked 4th for the prior 12 months, 7th for the prior 5 years, and 6th for the prior 10 years. Out of 50 industries, it ranked 4th for growth in profits. Only oil and gas equipment and services, homebuilders, and petroleum refining ranked higher.

Hospitals (medical facilities) faired only slightly worse. Their growth in profits ranked 6th. Only the metals industry stood between them and insurance companies.

Pharmaceuticals ranked 5th for the year 2005 for return on

investment as well as return on assets.

The entire listings are available in the April 17, 2006 issue of *Fortune* magazine.

The financial performance of these three industries is commanding. They rank near the top of all industries in the most economically powerful nation in the world. If the mandate of a corporation or company is to generate profit, why would any of these industries want change?

The answer is they don't. Regardless of the complicated schemes for reducing health care costs, regardless of the righteous indignation of those who are committed to free enterprise above all else, regardless of the enigmatic plans that involve tax credits, patient choice, and individual health accounts, the truth is, these three players are irrevocably committed to things remaining as they are. For them, change will never improve their financial performance beyond its present levels—with one exception.

That exception is a circumstance where one of the players gains an advantage over one of the others and benefits from a zero sum gain in profit. For example, if insurance companies can somehow gain leverage that allows them to reduce per diem hospital rates (the cost for one day of a patient's stay in the hospital for a given diagnosis), then insurance companies will experience increased profit because their overhead will decrease. This is zero sum because one party's gain is another party's loss, in this example the hospital company's.

That makes the "big players" opposing forces. What's good for one is always bad for at least one other. If prices for pharmaceuticals are forced down, it is good for hospitals and insurance companies and obviously bad for the drug companies. If pharmaceutical prices go up in the face of fixed insurance premiums, and it is time for renegotiation of hospital rates, you can bet that insurance companies will apply even greater pressure in those negotiations for lower hospital rates. (This recently happened in Denver and ended in a much publicized standoff.)

With regard to patients, this means that any significant health care change will always be opposed by at least one of the big players. Lower per diem hospital rates should obviously benefit patients by making health care cheaper. But any change within the health care system that would foster such change will be opposed by—you guessed it, hospital corporations. Insurance companies may sing the praise of such changes, but all the power of the American Hospital Association (AHA) will be brought to bear on making sure such change does not take place.

And let there be no doubt, these big players can apply leverage. In a special article appearing in the *American Journal of Medicine*, Landers and Sehgal pointed out that in 2000, health care lobbyists spent a total of $237 million, accounting for 15% of federal lobbying and representing an amount greater than the lobbying expenditures of every other sector. Small wonder that when the health care industry speaks the politicians listen. [2]

Further, as is graphically demonstrated in *Critical Condition*, one of the most popular jobs for former members of Congress is a lobbying job for the health care industry. Certainly, the most famous tie between our nation's capitol and the industry is that of Senator William Frist, whose family founded HCA, the nation's largest hospital chain. While Senator Frist has abandoned his presidential aspirations, he is at present still under investigation for the fashion by which he chose to divest himself of HCA stock. Since he was the Senate Majority Leader during the consideration of key medical legislation, this is not just an item of inconsequential muckraking.

With powerful players whose interests oppose each other and who are experiencing unprecedented financial success, is it any wonder that the so-called "incremental change" called for by the experts is in truth no change at all?

Voices in the wilderness

Just as I began writing this book, I read *Critical Condition*, written by Donald L. Barlett and James B. Steele. The scope and breadth of *Critical Condition* are so impressive that any attempt on my behalf to add further comment seemed silly. These fellows are Pulitzer Prize winning, well-armed investigative reporters who are as clearheaded as Carrie Nation.

---

[2] Health Care Lobbying in the United States, Landers MD, Steven H. and Sehgal MD, Ashwini R., Am J Med. 2004;474-477

If you have the slightest doubts about the origin of American health care's decline, read their book. It takes thousands of interviews and four years of research and uses them to paint a detailed picture of how the creation of a health care *industry* fostered medicine's decline. Their conclusion is that "Wall Street medicine" has not only sucked financial resources away from the goal of treating patients but has also eroded the traditional values of service, compassion, and professionalism.

When I first began reading *Critical Condition,* my family was a bit concerned with my behavior. After reading one particular passage, I leaped from my chair screaming, "You got it! Oh yes, you got it, got it, and got it! Oh yesyesyesyesyesyes -- hoohaw hoohaw." My wife was very upset, although I was never sure if her agitation was a sign of concern for my sanity or plain old anger because I had startled the dog into an acute problem with bladder control.

Critical Condition vindicated years of my howling at the moon—my writing of countless letters to the editors, angry e-mails, and vociferous magazine articles that were returned to me with red ink comments like "the editors think you should increase your medication" and "while I personally agree with you, I'd be insane to print this thing." Mr. Barlett and Mr. Steele have meticulously documented the unconscionable transformation of health care into a corporate profit center.

I briefly considered abandoning any further efforts at writing and instead simply buying hundreds of copies of *Critical*

Condition and handing them out. I'm sure Mr. Barlett and Mr. Steele would have appreciated such an effort, but a quick perusal of my personal finances revealed it would have been a short-lived campaign. Consequently you will have to go out and buy the book yourself. I strongly suggest that you do.

I decided to continue writing because the people who run the health care industry shushed quiet any significant response to *Critical Condition* by claiming that it was "inflammatory," lacking the "cool intellect" necessary for the appropriate changes needed in health care.

If *Critical Condition* is inflammatory, that is solely the result of the provocative nature of the truth. In fact, Barlett and Steele are somewhat reserved in their comments. What is needed is a work that truly *is* inflammatory because as things stand, *nothing* of significance is changing within the health care system. Worse, nothing *will* change without Americans becoming so inflamed that they demand it.

Barlett and Steele are winners of the Pulitzer Prize. I, however, have maintained my amateur status and am free to howl the truth in any way I wish. So I will. Let me clear my throat for a howl.

ANYONE WHO CLAIMS THAT OUR MARKET DRIVEN HEALTH CARE SYSTEM IS NOT WORKING IS *STUPID*. IT'S WORKING JUST THE WAY IT'S SUPPOSED TO, AS A *PROFIT GENERATING INDUSTRY*.

A market-driven system is designed to generate *profit* for

America's corporations, and that is exactly what our system is doing. It's successful because its product can be hawked with a unique sales pitch, "If you don't buy this you might *die*." Now there's a product! No wonder this industry is so successful. No wonder thirty years ago, when a government economist with the common sense of Daffy Duck decided that the only thing medicine needed was marketplace competition to drive down costs and increase quality, Wall Street responded like an eighteen-year-old male after an overdose of Viagra. They were excited with a lust that had nothing to do with long-term goals, love, or any virtue that was not associated with an immediate piece of the action.

Now, when the results of that action have gestated, so to speak, and the effects of the Viagra have worn off, the experts are feigning surprise and the Wall Street lovers of humankind are clucking their tongues about how much money is being spent on the medical industry.

Market driven systems don't *decrease* utilization of their goods and services. They *increase* utilization. Market driven industries are designed to make *profit*. That's the primary goal of a market driven system. Anything else is secondary.

When you see a television ad for a drug, does the ad say: "*Curesitall,* the little pink pill that will stop your acid problem, give you the special strength to satisfy the little lady, and drop your cholesterol so low that when you shake hands with your neighbor, you'll drop *his* cholesterol twenty points. *Curesitall,* the drug that cost more to develop than it cost to put a man on the moon, but don't use it. We've paid six million dollars for this Super Bowl halftime ad to tell you to use this drug only if it's absolutely necessary! It costs as much as solid platinum, and you probably don't really need it"?

No. The ad does not say that. The ad shows an actor and actress who start off looking like they just survived the bombing of Dresden, take *Curesitall,* and seconds later look like they're on their way to that Chinese place in Hollywood to pick up a couple of Oscars.

Does the spokesperson for an insurance company call a press conference and say, "We've been looking at the cost of health care and damn, it is going through the roof. Because of this, we are going to cut the profit margin on our policy premiums to try to impart a bit of financial restraint on this mess?"

No, no, no, no, no. That's not what he says. He walks up to the podium and mutters long enigmatic phrases about "the natural cycle of insurance premiums" and "unfortunate trends in an aging population" and concludes by announcing, "We therefore find it necessary to increase premiums by eighteen percnt and decrease services by a corresponding amount." He turns *off* the microphone, covers his mouth with a handkerchief and whispers "our profits increased last year by thirty-eight percent." He coughs, puts the handkerchief in his pocket, and backpedals off the stage like a NFL cornerback.

119

Medical Metamorphosis

The market driven medical industry stands as much a chance of curtailing costs as the United States Congress does of eliminating lobbying. Neither of them is designed for that sort of behavior. I don't care how you package it, how it's reconfigured, it cannot control costs. Whether an employer gets a tax credit or even if every patient must pass a test proving he is an informed consumer before he's allowed to officially *have* a disease, it will never, never do what the experts are trying to make it do. As long as the American health care system exists as a *pure* market driven medical industry, costs will continue to *increase.*

This is not an indictment of profit or a criticism of capitalism. It is a simple statement of fact about a commodity for whose demand there is no end. Death, suffering, aging, pain, and injury are an eternal part of the human condition.

Even the ubiquitous demands for energy and food can be modified to some degree. Thermostats can be turned down and sweaters worn, driving vacations can become short trips, and fresh avocado salads can be replaced by lettuce and tomatoes. But when a person is in pain, is frightened, or fears for the life of a sick child, that person will pay any price to stop the pain, end the fear, or make the child well.

In that setting, the consumer—the patient—is at the mercy of the vendor. If the vendor adheres to the free market tenet, "whatever the market will bear," then the market will have to bear a heavy load.

Madness is badness of spirit, when one seeks profit from all sources.

--Aristotle

CHAPTER EIGHT

GLUTTONY

Greed personified

Critical Condition documents in detail how each of the players has engaged in self-service and pushed America into the street wearing oversized pants and undersized coats, stumbling on its way. For me to retell the stories and recount the figures that Barlett and Steele have meticulously assembled would be silly.

But one player deserves particular attention.

Neglect, indifference, and uncertainty—greed, hypocrisy, and irrationality changed our health care system. We all share the burden of neglect, indifference, and uncertainty, but some share a glutton's portion of the burden of greed, hypocrisy, and irrationality. The insurance industry is the fattest of the fat.

Prior to 1970, most insurance companies were non-profit. That oversight was remedied as Wall Street snuggled up to insurance regulators and whispered into their ears the sweet nothings of financial incentives, mammoth fees, and new economic possibilities. There then began the transformation of the Blue Cross, Blue Shields of America into profit centers.

That transformation continues to this day. The few non-profit insurance companies that remain are being regarded with ravenous intent. One of those companies is Premera, located in Washington State. Since 2003, its executive officers have been petitioning the Washington State Insurance Commission to convert from a non-profit to a for-profit company. The insurance commissioner has made a decision that such a conversion would not be in the best interests of the citizens of Washington. The officers of Premera are appealing that decision.

Premera's officers have defended their efforts in countless press releases. In response, their motives have been questioned by a number of citizens groups. The Premera Watch Coalition released a report entitled "Who Benefits? The Role of Executive Compensation in Health care Conversions." I doubt the title would win any awards for subtlety, but the Coalition has no desire to be subtle. Its members think Premera's executives are just plain greedy.

Arithmetic would appear to support the conclusion that if not greedy, they are at least deeply steeped in self-interest. I use the word arithmetic in a very specific sense. It is the simplest mathematical discipline I know. Common sense and simple logic are all that appear to be required to come to an indisputable conclusion about motives for an insurance company's conversion to a for-profit status.

Very simply, what Premera wants to do is to issue stock and then operate the insurance company with a goal of bringing in more money from insurance premiums than the company pays out for patients' health care. The difference is profit. At present, given that Premera is not-for-profit, that difference is treated as either a reserve to be held for future medical catastrophes like a bad flu season or used to reduce patients' premiums. Premera is suggesting that it is better to pay dividends to stockholders than it is to create a reserve or reduce premiums. It's really no more complicated than that. Simple arithmetic proves that this means health care will be more expensive because at some point there *will* be a bad flu season, and the reserves that have been used to pay stockholders will have to be made up by increasing premiums. If the money paid to stockholders would have been used to reduce premiums rather than held as a reserve, then—listen up economists—health care costs would have been lower because premiums would have been lower. This is not calculus or complicated statistical analysis. It's arithmetic. An insurance premium that must pay stockholders and pay for health care *must* be higher than one that just pays for health care.

Premera would argue that my explanation ignores capitalization—the money stockholders initially pay for stock. There are some who would argue that the only reason a company should sell stock, "go public," is if it needs money—right now. A stock issue is a means of raising capital instead of taking a loan, or robbing a large number of convenience stores, or collecting an even larger number of aluminum cans. Thus, if Premera needs immediate cash and doesn't want to consider loans, cans, or robbery, then converting to for-profit makes sense.

One of Premera's press releases states, "Like other health insurers, Premera needs capital to increase insurance reserves in response to rising medical care costs; to invest in better technology and services for members..." So, there it is. There is the reason the executives want to issue stock. They're not greedy. Premera is strapped for cash. They need to create those reserves for that rainy, fluish day in the future.

Oops. The only problem with this line of reasoning is that it's pure fabrication. To quote the Sunday December 13, 2005, *Seattle Times*, " The three big health insurers in Washington— Premera, Regence and Group Health—have been quietly piling up more money than they need... from 2000 to 2004, capital and surplus at Premera and its subsidiary, Lifeline, increased from $271 million to $446 million."

Arithmetic prevails. Premera needs capitalization about as much as Washington State's premier citizen, Bill Gates, needs a part-time job. It's flush with cash. Why then would Premera's executives want to issue stock and ...wait a minute, is it possible the Premera Watch Coalition is correct? Could these executives really, truly be greedy?

There is some dispute about exactly what a Premera conversion would mean for its officers. However, the Coalition's report cites financial statistics from a number of other conversions from non-profit to for-profit. Executives of these companies enjoyed an increase in remuneration ranging from 1,035% to 1,505% of their pre-conversion salaries. In gross dollars, for example, Anthem Insurance Company's CEO/President Larry Glasscock was paid $15.9 million in total compensation in 2002.

Nice bonus.

The numbers game

Once a conversion to a for-profit status has been made, the basic goal of an insurance company changes. Like any business, for-profit insurance companies have a responsibility to their consumers, but unlike non-profit companies they additionally carry the heavy burden of responsibility to stockholders. How well an insurance company can carry this new burden determines the success of the company's stock. How well the company's stock does determines the ongoing salaries and bonuses of company executives.

As the cost of medical services increases, insurance premiums obviously increase as well. But insurance premiums increase quicker than the actual cost of health care. This difference is that "heavy burden," the profit, that so excited Wall Street when health insurance became a for-profit industry. It is the "responsibility to stockholders" that must be generated over and above its operating costs.

As we have already discovered, insurance companies are

doing quite nicely in their efforts to meet these responsibilities. Most of America is pleading for relief from the cost of health care, but recently, a spokesman for the health insurance industry described the success of HMO's as "knocking them dead." Only the spokesman knows whether this particular choice of words was intentional.

Since the high cost of premiums has deprived a large segment of Americans from carrying health insurance, since many will not seek medical care without insurance, since not seeking medical care can easily have untoward medical consequences, including those that prove to be fatal, the phrase "knocking them dead" is literally true as well as a figurative description of these nice, fat earnings.

The conversion to a for-profit status adds nothing to the care of patients other than a surcharge on every dollar spent for health care. This surcharge is the profit portion of the arithmetic that makes an insurance company a certified member of Wall Street medicine. Premera and its brethren dispute this by creating various charitable programs and adding clever patient benefits. But the function of an insurance company is to spread risk over a large population. That is their mandate, their raison d'etre—not to hire nurses to call diabetic patients once a month and then declare these token telephone calls a program of quality assurance.

Nancy O'Connor of CodeBlueNow, a Washington State organization that supports citizens' health care rights, points out that Premera's record regarding a commitment to the state of Washington is already somewhat tarnished. "...Premera is dropping Medicaid, Basic Health Plan and state employees—and playing hardball with Multicare and Providence health care—what makes anyone think it cares about the marketplace, the businesses or the people in Washington and Alaska?"

The American health care system cannot afford its resources to be consumed by entities that do not enhance patient care. For-profit insurance has no redeeming virtues. The pharmaceutical companies have taken heat because of their amount of profit, but no one can honestly discount the benefits of their discovery of miracle drugs like the statins, medications that reduce the incidence of heart disease, Alzheimer's disease, and sudden death syndrome—three rather nasty propositions.

I have yet to encounter anyone who can even *suggest* a similar benefit from for-profit medical insurance. The only thing it has done is make some people a whole lot of money. Our friendly CEO at Anthem did a bit better after his meager $15.9 million income in 2002. When Anthem and Wellpoint merged, he received a $58 million bonus. Every penny of that money has been taken away from funds that should go towards treating patients.

Far from enhancing patient care, for-profit health insurance adds to its problems. Administrative costs now account for *greater than one out of every three health care dollars.* This is triple the amount of money spent on administration in Canada.

127

Insurance companies also add to the inefficiency of medical care in ways that are difficult to quantitate and therefore don't show up in financial statistics. When a physician spends time pleading for the right to treat a patient, that time is lost. It is a real loss of a valuable resource no different from flushing expensive medications down the toilet.

The experts rarely criticize the parasitic nature of for-profit health insurance. Their attention is more often focused on issues like the need to establish new mechanisms for deciding whether or not particular medical treatments are efficient and effective; for example, studies that evaluate whether a new expensive blood pressure medication is any better than an older generic.

I have no problem with any process that adds to our understanding of medications and their advantages or disadvantages. But this should be an empiric process, apart from jostling imparted by economically interested parties. Evaluating any medical treatment is tricky even when done in an environment devoid of hidden agendas. Further, any beneficial financial impact of such processes is miniscule when compared to the detrimental impact of for-profit health insurance, an industry that takes *a minimum* of *thirty-three cents* off the top of each health care dollar and offers nothing in return.

In my business this would be like saying to a patient, "Well Bob, that truck that hit you managed to break your right femur, your left tibia and fibula, your left humerus, and four ribs.

You've also got a collapsed lung and a significant pericardial contusion. Holy Moly, I'll bet all that stuff smarts.

"But Bob, I was looking at your electronic medical record, and it appears that you've not had a good prostate exam in some time. You know, Bob, you're in that age group that needs to be checked annually—the data is pretty clear on that front. So, what I'd like you to do is stand up, bend over, and I'll do a rectal exam."

Quite obviously, after asking for a second opinion and some morphine, Bob should also tell this clueless doctor to go do a rectal exam on himself. While not being so socially maladjusted as to suggest something similar to health care experts, I will suggest that they consider their priorities and apply a bit of common sense. Health care is hemorrhaging dollars from a gaping wound inflicted by for-profit insurance. This wound needs much more immediate attention than a pimple, even if the pimple is on the end of the patient's nose.

I am not condemning a free market economy. I think a free market economy is one of humanity's greatest inventions. It usually *does* foster competition and usually *will* increase quality and decrease cost. But with regard to health care, it has failed miserably to do either. When an industry deals in commodities controlling life or death, those commodities are not subject to typical market forces—for all the reasons we have discussed.

Even those with an ability to pay for health insurance are often unable to obtain it. Insurance companies are aggressively trying to cover the healthiest Americans while ignoring those with health problems. If you have a chronic disease, just try to get insurance. Insurance companies want to insure people who do not need to be insured, not those who have chronic diseases.

That's hardly a revelation. Insuring people who are healthy is a necessary part of health insurance because without them, there would only be sick people paying premiums. That would mean that sick people would have to pay enough premiums to pay for all the sick people. That means premiums would be outrageously high.

Insurance companies take in money in the form of premiums and pay it out in the form of medical services. Period. Nothing magic happens when the premiums reach the insurance companies. They do not grow—at least they are not *supposed to grow*. Unlike premiums on life insurance that are held for many years and only paid out when a person croaks, health care premiums should be paid out ASAP, in the form of services.

Health care premiums should be used to take care of patients, not loaned out to build large buildings.

Consequently, premiums from healthy people, those who are unlikely to use medical services, are needed to offset the costs of sick people. Otherwise, insurance companies would not be needed. Healthy people would keep their money, and sick people would get creamed.

The entire reason for insurance is to insure that you don't get creamed. The assumption is that people will be willing to pay into a common fund knowing that someday they *could* get creamed because if they are lucky they will live to be old enough to develop the problems that old people get, and if they are unlucky, they will fall off their ten-speed bicycle, break a leg and need insurance when they're young.

That's insurance in a nutshell. It's not rocket science. The system breaks down if young healthy people think they are bullet-proof and would rather buy a new ten-speed bicycle than pay for insurance. It also breaks down if insurance companies act like any other profit driven corporation and realize that young bullet-proof people won't buy health insurance so they make profit the only other way possible. They reduce what they have to pay out in the form of services.

How do insurance companies do that? By *rationing* medical services and by *not covering sick people.* There is no other way for-profit insurance companies can be profitable. None. Nunca. Rien, Zero.

If insurance companies are profitable and young people think they are bullet-proof—both of which are true—then the rest of America is screwed. I mention again, this is not rocket science. This is arithmetic and reality.

Medical Metamorphosis

Truth, like light, is blinding. Lies, on the other hand, are a beautiful dusk, which enhances the value of each object.

-- Albert Camus in *The Fall*, p. 126, Gallimard (1956).

CHAPTER NINE

INCREMENTAL CHANGE IS NO CHANGE AT ALL

A fog thicker than ~~sea poop~~ pea soup

It's now time to take a look at the suits themselves—the ill-fitting suits that America has been asked to wear. These suits all wear a single label—*incremental change.* Unfortunately for America, this label is being passed off as Gucci or Yves St. Lauren.

Incremental change contains elements that might best be described as a product of magical thinking. What else could foster suggestions that patients can take care of themselves if they are well enough educated; that electronic medical records, by themselves, will significantly improve the quality, cost, and efficiency of medical care; that the nursing crisis can be solved by a federal program of television commercials changing the public's perception of the nursing profession; and that the shortage of physicians predicted for the next decade won't actually be a problem because the general health of the public will improve? These have all been suggested as parts of an overall plan to gradually improve health care.

In other words, America doesn't have to worry about a shortage of doctors or nurses because even though the population is aging so fast that there will not be enough young people in the working force to support Social Security, even though obesity is reported as being epidemic, and even though our lifestyles are characterized as being self destructive, the population is becoming healthier. Besides, we won't need doctors because patients will be taking care of themselves, because they will learn medicine on the Internet. The nursing shortage will be solved when lots of people choose to spend their lives working in the nursing profession just because they saw a public service advertisement at two in the morning, even though nurses are underpaid and forced to work in conditions where their workload is so great that they make life-threatening mistakes. And even if there might not be enough doctors or nurses, they're not nearly as important as the way they keep records of what they did, and that's going to change dramatically once someone decides who's going to pay for converting to electronic medical records, and how to make a records system that's as big as the health care system actually work, and agrees how to choose a system, and how to do all that while complying with the HIPAA act that basically says it's illegal to do any of that. See? No problem.

Are you kidding me? If that all sounds a bit silly, it should. It *is* silly. It is puffery pitched by hawkers of snake oil that is guaranteed to cure the pains of a gangrenous system of medical care. (I'm rather proud of that sentence—it's not a bad piece of pitching itself.) The problem with this fog of deception is that it hides the true nature of what is being suggested under the guise of incremental change.

Health Savings Accounts

Let's take a specific look at a number of ill-fitting suits and begin with Health Savings Accounts (HSAs). HSAs, also called Individual Health Accounts, are touted as offering "freedom of choice." They are also hawked as a means of forcing patients to consider the economic ramifications of medical decisions—"It's always easy to spend money when it's someone else's."

In truth, they accomplish two goals. The first is cost shifting to patients. The ability of state and federal governments and employers to fund health care has reached a breaking point. The only other source of funding is individual Americans. HSAs are a means of doing exactly that. Period.

Any fiddling with tax credits, any promotion that suggests this will save patients money in the long run is rubbish. Tax credits may offer a means of softening the blow to some parts of the population, but they are designed to shift the *overall* financial burden of increasing cost to the backs of patients. Period. Patients may save money in the long run, but only when compared to an internal reality where some form of HSAs are already in place. In

135

other words, on Monday, HSAs are proposed. On Wednesday, a slightly different form of HSA is proposed that indeed saves money in the long run but only if compared *to the first HSA proposed on Monday*. It's a classic "bait and switch." Period.

Worse, it is also part of patched-together programs proposed by the governors of a number of states under the guise of achieving universal coverage. Most of these programs are centered on the demand that all residents of a state have health insurance of some sort, much the same way all drivers are supposed to have auto insurance. This means that those who think themselves bullet-proof will still be forced to buy health insurance.

The means of offering affordable insurance are achieved by pushing HSAs, hammerlocking insurance companies to offer catastrophic insurance plans that are priced at a lower premium by offering government assistance, and offering tax breaks to companies offering insurance programs as a benefit. The goal of these programs is for the state to be able to claim 100% coverage.

But consider the details of some of the policies. The most egregious are those offering catastrophic coverage. For a family of four, one of these programs located in California was quoted as costing $2,400 per year. That is $200 per month. This may not seem like much, but if the total income for the family of four is $29,500, that may exceed any amount allotted for discretionary spending.

Further, what the family has done is place themselves at

risk of a financial catastrophe, not prevent one. These policies will almost always have a deductible in the range of between $2,000 and $5,000. After the deductible has been met, the policies will usually pay at a rate of 80% for the first $20,000 to $40,000.

$40,000, the upper part of that range, is not an unusual total for a hospitalization that includes surgery. Let's say Jimmy falls off his bicycle and incurs a badly fractured leg. He's hospitalized, has his leg operated on, and is in the hospital for five days. Let's say the deductible is $5,000. His parents' bill for the hospitalization would be $5,000 + $7,000 or $12,000.

A bill of $40,000 for this family would have been catastrophic and would have forced them to consider something like bankruptcy—even with the new bankruptcy laws. That's bad. But what's worse? A bill of $12,000. It's just as deadly financially, but it's small enough that bill collectors could attack wages, attach liens, and use all the means that are used when an *almost* lethal financial blow is administered to a family.

Catastrophic health insurance policies are great for young, healthy, high-flying executives who can handle $12,000 and want to avoid the exposure to hits of $40,000. They are catastrophic for families on the edge.

They are also great for hospitals and vendors of high-priced medical services. Why? Because *they get paid.* The $28,000 that the insurance company pays out would not have been paid had the family simply carried no insurance at all.

137

Catastrophic medical insurance does nothing for America's working lower middle class and below. It is purely a way of protecting the "players." Families who carry such insurance may very well help a governor's batting average in an attempt to claim "universal coverage," but they are in truth worse off than if they had no insurance at all.

The second goal of HSAs is protection of for-profit insurance. When all is said and done, whatever money is salted away in an HSA must be used to buy insurance from the helter skelter disarray of insurance companies that exist throughout America. Each insurance company has its own rules. Each company has its means of maximizing profits. These companies will continue to operate in a fashion that generates those profits with even greater ease in the face of HSAs because the heat will be off.

HSAs do nothing to increase efficiency because they will do nothing to effectively increase competition. There has already been competition applied by employers. That's why patients are forced to change doctors on an almost annual basis. Their employers have changed insurance carriers to get "the best deal." Has this slowed the increase in health care costs?

If it had, HSAs would not be needed. Think about it. How much leverage does some poor self-employed patient have with his HSA when compared to an employer who may have thousands of employees? In a market where health insurance products are more complicated than a textbook on quantum mechanics, how "educated" a consumer is the isolated patient when compared to the resources available to General Motors? If General Motors can't control what it spends on health care, are we to believe that John Q. Public, armed with his HSA, is going to control it?

As might be predicted, the state run programs are having problems even early on into their implementation. The basic problems stem from the bullet-proof population not buying insurance and the insurance companies doing their bureaucratic best to dodge sick people, ration services, and still make a profit.

In other words, the mandate that there be universal coverage is being ignored.

The states are having problems because mandating "voluntary" participation is what is called an oxymoron—"a figure of speech in which opposite or contradictory ideas or terms are combined" (Webster's New World Dictionary). If it's a mandate then its not voluntary, and anything not voluntary must have some force making it mandatory, and the state is a bit reluctant to put people in jail for not having health insurance.

If they don't put people in jail, then how do they threaten them? Fines? But if the fines are not paid, what happens? Jail? And who collects the fines? And how much should the fines be? If they are less than the premiums for health insurance, then why pay the premium? If you discover you're not bullet-proof, why not just start paying the premium at that time?

I have a headache.

Medical Metamorphosis

Whoever places his trust into a system will soon be without a home. While you are building your third story, the two lower ones have already been dismantled.

--Franz Grillparzer in "Vischer's Aesthetics," Poems (1858).

CHAPTER TEN

SYSTEMS

Systems versus people

The next ill-fitting suit is systems. Those running the medical industry deal in abstractions. An applied abstraction becomes a system. Experts worship systems. It is their belief that all the problems in health care can be solved through the use of systems. There is an implicit assumption that systems of operations, systems of data storage, and systems organizing systems are more important than those who must implement those systems.

As an example, let's return our attention to Dan, our patient with chest pain. The experts might look at his case and observe:

1. The doctor should have had assistance in following up with his patients. A team approach to patient care would

relegate initial follow-up to his medical assistant, a nurse practitioner, or physician's assistant, thus utilizing the physician's time more appropriately.

2. In a circumstance as serious as chest pain or a possible suicide, there should be a system in place such that these patients' charts are red flagged until an appropriate outcome is achieved.

3. The doctor should not be seeing patients on the morning following a night on call. Office hours should begin later in the day such that no patient suffers because of physician fatigue.

At face value, these suggestions would appear to be insightful and an appropriate solution to the problem. Of course they are. My partners and I thought of them while spending thirty years trying to design systems for our practice. I am, after all, no dummy. Who wants to spend his life running around like his hair is on fire?

This follow-up system appears completely functional—on paper. In reality, it consistently failed because we were always short staffed, a common problem for primary care offices. In addition staff cuts caused by lower reimbursements, insurance companies and the government have placed ever increasing administrative burdens on primary care providers, and these burdens have shifted the efforts of employees away from patient care. Because primary care doctors are unable to match the wages and benefits of corporations and higher-paid specialists, their offices also have a high employee turnover rate, with often the best employees chasing a better paycheck.

A system that is designed for four employees will not work when only three are available. In a primary care office, patient load is unpredictable and emergencies can pop up at any time. What evolves is a work environment centered on daily battles to stamp out fires while still trying to meet patient demands. Experts can pontificate until they run out of wind, but that's just the way it is in most primary care offices. The best designed systems fail miserably because their financial base erodes as the larger, more powerful system of insurance companies gobbles up dollars, and because Mother Nature refuses to make primary care predictable. ("I'm sorry Mrs. Brown, but little Tommy is not scheduled to fall off his bicycle until next week. You'll have to wait until then to have that gaping, bloody wound on his lip taken care of.")

I learned early on that the people with whom I worked were infinitely more important than was our system of operations. *No system is any better than the work of the people who must implement it.* Well-designed systems utilizing the most up-to-date technology can be powerful tools, but that is all they are—tools. Without a dedicated, well-trained professional using these tools in the best interests of a patient, they are no more than hammer and nails without a carpenter. Worse, the medical industry is so enamored with these tools that it ignores the importance of the carpenter altogether. The experts view doctors

143

and other medical professionals as more of a nuisance than a crucial part of health care. I once read a full-page newspaper article dealing with changes in health care and how they might impact cost that *never* used the word "physician."

The medical industry glorifies systems, but it abuses those who must implement them. Front-line health care providers—primary care physicians, hospital ward nurses, emergency room physicians, emergency room nurses, x-ray technicians, emergency room admitting clerks, etc.—are constantly overburdened, underpaid, underappreciated, scapegoated, and faced with never-ending demands that are impossible to meet and tasks that can never be completed. Consequently, they can become a weak link in the health care delivery system when they should be its strength.

For a doctor, abuse at the hands of the system is best exemplified by a situation that has become a cliché—spending time on a telephone arguing with an insurance clerk about whether a patient may have a medical procedure. The clerk's job description, written by an accountant or the equivalent, is to act as a hindrance to the procedure's authorization. The accountant's assumption is that the insurance company's bottom line will be greater paying a minimum wage clerk to delay or prevent medical services than making their authorization efficient. If this roadblock completely stops only two or three procedures a month, the clerk's salary is effectively paid for, and the insurance company's bottom line goes up. Further, and I have been told this personally, there is an assumption that if the procedure really needs to be done, the patient's physician will be persistent and eventually succeed in having it authorized.

The accountant has not considered the toll this process takes on the physician's time because the accountant does not care. There has been a net decrease in the insurance company's overhead, and that is the entire goal of the system. That a hurdle has been placed in front of a professional trying to care for a patient is of no import to the accountant or the insurance company. Economists or insurance executives who deny this are quite simply lying through their porcelain caps.

Medicine is not unique in this regard. Chances are, unless you're comfortably situated at the top of the food chain, you too are overburdened, you too are underpaid etc., etc., etc. America is now a bottom-line proposition. Policies are established by accountants instead of skilled, trained, and proficient employees and professionals. The religion of the bottom line, wherein profit is god, demands that those forced to worship at its alter and still appropriately serve society must often do heroic work. I see these heroes every day.

One of my heroes

I am a convenience store person. I begin every day stopping at my nearest 24-hour convenience store, purchasing coffee and a doughnut, spending a few moments in conversation with the employees, and then leaving for wherever I'm going. As

Medical Metamorphosis

I drive there, I distribute between five and ten percent of the coffee and doughnut on my clothes and consume the rest.

I have been going to the same convenience store for more than twenty-five years. I am sure that places me in some category scorned by culinary experts, sociologists, and gastroenterologists, but it does makes me an expert in the history of this particular convenience store. I have come to know a great many clerks from a great many countries. I have also witnessed the careers of my store's managers last for periods as short as three days. Until its present manager, the longest of these careers was four months. The present manager has been there more than ten years. She is one of my heroes.

I will call her Nora, and Nora is loyal, tough, courageous, unbelievably hardworking, creative, and personable. She treats her employees, who come and go with great rapidity, with respect and a truly caring dignity. She has faced down kidnappers and shoplifters. She has returned to work three days after almost being killed in a hit-and-run accident, and until recently, I never heard her complain. In the last few months, she has finally begun acknowledging how little she is respected by the large corporation that owns the convenience store. Her frustration with the "system" has finally overwhelmed her ability to quietly do her job. Now, she acknowledges the insanity of many of the system's demands. With those of us who are regulars, she will also admit that she is paid a pittance compared to the difficulty of her job.

I once asked her if she had any desire to do something else. The familiarity of her response was painful. "Oh no," she said, "I love my job. I love the people who come in, who depend upon us to deliver the simple things that help them get along. Hard work doesn't bother me. I grew up on a ranch. I've worked for the company for twenty-three years. Only two years until I can retire. It's just the..."

"The bullshit," I said without thinking.

She smiled. "You got it. The bullshit."

In my opinion, a really important, high-placed, company muckety-muck ought to be personally interviewing Nora, begging her to favor him with her knowledge and wisdom. Then, just after he gives her a big bonus, he ought to be pinning a medal on her little red vest while a very large band plays a Sousa march. For crying out loud, until Nora arrived, there were literally nights when the store was closed because the manager walked off without even giving notice. Before Nora, the condition of the store varied from okay to disorganized, from kinda dirty to disgusting. This went on for twenty years. Then Nora arrived. My convenience store is now a thing of beauty.

There will be no bands or medals. Nora's skills and dedication continue *in spite* of, not *because of,* the system. To hell with a system. Simply hire lots of Noras. The accountants will respond to that suggestion by claiming there aren't many Noras. My experience tells me otherwise. It also tells me that the religion of the bottom line destroys those who might evolve into

147

Noras.

If you are an average American working for someone, you're probably not often reminded of your value. Maybe your boss at a company Christmas party says something after he has knocked down his sixth glass of holiday punch. You may have also been at a meeting when management began by praising you and ended by announcing either a pay cut or a downsizing. In my profession it is most hypocritically professed by insurance executives, "You doctors are the heart and soul of our health care network. We know how much you mean to our subscribers, our patients. We just love you guys and gals, and we just love our patients. Yada, yada, yada.....yada, yada, yada. And in closing we are *increasing* our health care premiums by 18% and *reducing* your reimbursement for office calls by 15%." (In the Army, this is venerated in the acronym BOHICA—Bend Over Here It Comes Again.)

There are obviously exceptions to this generalization, but it is the generalization that now has the power to make changes in the health care system. It is the generalization that values systems over people, even though elemental reasoning and history decry such an assessment of value.

The war on disease and suffering

Medicine has been called a war against disease and suffering. I've found such a metaphor to be both utilitarian and amusing. In that regard, there is a dramatic historical example of the comparative role that systems and people play in humanity's destiny. In terms of scale, complexity, and the importance of what depended upon its outcome, the Allied Force's invasion of Europe in 1944 was humanity's greatest undertaking in the last one hundred years. In retrospect, the transportation and coordination of massive amounts of equipment and huge numbers of personnel, the design and manufacture of the equipment, the integration of different cultures and nationalities, the acquisition and evaluation of data describing the capabilities and probable future behavior of an enemy trying to obscure both, and the effective concealment of all of this from the enemy is almost incomprehensible in its magnitude. Obviously, the concomitant systems were detailed, created by the best minds available *anywhere,* and virtually unlimited in their access to resources. D-day's comptroller had what you might call an ideal budget—no top end.

The people involved with the invasion were well trained and motivated, but since we are talking about hundreds of thousands of people, most of them, were just average folks trying to do a difficult job.

So, here is an historical venue where we can assess relative value, people versus systems. Which of these two do you suppose saved the day? Was it the very best systems with unlimited resources? Or, was it the underdog, the run-of-the-mill human being plopped down in the middle of the twentieth century's big showdown of good versus evil?

Medical Metamorphosis

People won, hands down, going away. They were dropped from airplanes in the wrong place; their airplanes bombed the wrong areas; they watched their comrades die in scores right next to them; innovative inventions like floating tanks sank beneath them; they were so overloaded with equipment that many of them drowned before they even set foot upon the continent they were invading; chains of command were destroyed as their officers became casualties; battle plans had to be abandoned when some of them met little resistance and others measured progress in yards. But the Allies prevailed because the soldiers fighting the battle innovated, used their skills, reacted to unexpected circumstances, did jobs not technically in their "job descriptions," and distinguished themselves with acts of courage, dogged determination, and unimaginable hard work.

Axis soldier's fought bravely but another truth about systems predestined their failure: *No system can be ultimately successful if it limits the skill, creativity, or initiative of those who must implement it.* The German chain of command was a classic example of rigid, top to bottom, insanity—literally and figuratively. No one did much of anything without permission from the top. Unfortunately for the German armed forces that position happened to be held by a little fellow who suffered from megalomania of the worst sort, severe drug addiction, dementia from advanced Parkinson's disease, a lousy disposition without the saving grace of a sense of humor, and a truly bad idea of what comprised a good-looking mustache.

In his book *D-Day*, the celebrated historian Steven Ambrose addressed the issue of systems and plans and whether they carried the day at Normandy. "It was not a miracle. It was the infantry. The plan called for the air and naval bombardments, followed by tanks and dozers to blast a path... but the plan...failed, utterly and completely... As is always the case in war, it was up to the infantry."

The reality of a military war dictates that plans and systems *must* fail because war is a rather unpredictable proposition. Weather changes, the enemy does not behave as predicted, ("Damn those Colonists. They're actually hiding behind trees and shooting at us!"), equipment fails, and people's behavior runs the gamut from acts of courage to the self-preservation of cowardice. The war on disease and suffering is equally capricious. Patients and their maladies are almost whimsical in the way they present; the stress of an illness can produce unique and unexpected emotional reactions; and the degree of urgency required to avert a disaster is often obscured ("You're kidding! A heart attack? And here for a whole week I thought I had heartburn. Pretty funny, eh Doc? Doc? Uh, Doc, you're not laughing. We got a problem?"). There are also a seemingly infinite number of ways a human being can get fouled up and a stunning array of ways patients screw up their treatments ("That was a *rectal* suppository? I thought it was just a really big pill."). Only a well-trained, creative, adaptable member of the infantry in both of these wars has the capacity to evaluate, react, and cope with such randomness. No system or plan can be designed to anticipate its scope.

Systems can, however, destroy any chance for success. Following WWII, I think it fair to say most of the surviving German officers agreed it might not have been a good idea having a lunatic at the top of a rigid chain of command. Those running health care may not be lunatics, but they are certainly bereft of any understanding of medicine's subtle realities.

Systems—the myth of electronic medical records

We've already taken a look at the importance of an accurate medical history. Incremental change ignores this importance. But there is no better example of an unfailing devotion to systems than electronic health records (EHR)—the way histories—accurate or inaccurate—are stored

Basically, EHR is simply the use of a computer to store medical records. It's substituting the word processor for pen and paper and a spreadsheet for hand-plotted graphs. However, the universal adoption of EHR is touted as a way to reduce medical errors, cut the cost of health care, assure the quality of medical care, educate patients, eliminate all dangerous drug interactions, and coordinate care among specialists.

It is true that taken individually, these tasks can all be *facilitated* by EHR. Handwritten notes and orders are often illegible; programs for the analysis of data can be "tagged" to the chart so that a clinician is more likely to notice trends of significance; and the medical chart itself, a physical document often separated from a patient, becomes instantly available "at the stroke of a key" instead of being misfiled or sitting in a six-foot pile of charts on a doctor's desk. Theoretically a record stored as EHR will also contain all data, lab reports, and results of x-rays because they can be immediately and magically zapped from the computer of their source to the computer containing the patient's record.

Thirty years ago, when this sort of data management was the stuff of science fiction, my response to its possibility was "cool" (Yes, I know that dates me). Now, given the fact that EHR has become health care's panacea, my response is, "Are you kidding?"

First of all, the idea that America should focus its attention on eliminating pen and paper instead of the fact that 45 million Americans have no health insurance is an insane distortion of priorities. Secondly, billing EHR as a cure-all is the ultimate "spin" on the reality of health care's crisis. It's a way for those running the show to abdicate responsibility for making basic—and politically painful—changes, saying instead, "the medical community just needs to come out of the Stone Age, get rid of their Mont Blancs, and everything will be fine."

Rubbish.

A national system of EHR is also impractical to a point of impossibility. Ironically the bureaucrats who are demanding such a system are the same ones who have made it impossible.

153

To begin with, it is widely accepted that such a system will cost billions of dollars. The federal government has no intention of investing those dollars. Records from the offices of PCPs are a crucial part of a universal system, but most PCPs consider the cost of EHR prohibitive. Reimbursement for primary care has dropped to such a degree that they have had to fire staff, take large pay cuts, and in some cases close their practices. Ask an average PCP what he or she thinks about investing tens of thousands of dollars in a system of EHR.

Then duck.

Hospitals and insurance companies would benefit directly from EHR and in some areas *have* invested in software and equipment. However, to close the loop of an EHR system, doctors' offices and their outpatient records must be included, and this happens to be illegal.

There is a federal law called the Stark Law that prohibits physicians from any gain because of referrals to labs, hospitals, medical facilities, or other physicians. This law has been called the "kickback law." In theory, it legislates against physicians having financial incentives to use a particular specialist or medical facility, i.e., financial gain is not supposed to play a part in decisions about medical treatment. Nice theory, but in practice this law often hurts patients.

It's worth a parenthetical digression to explain what I mean. Consider this example from my own experience. The Stark Law prohibited doctors in our medical building from owning an x-ray unit. With an x-ray unit in the building, my 86-year-old patient with a possible leg fracture would have seen me, been wheelchaired by my medical assistant to the x-ray unit where she rarely would have had to wait for more than a few minutes, had the x-ray taken, and would have brought the films back for me to review personally. The patient had to travel only the distance to and from the elevator. During the entirety of her trip, she would have been in a wheel chair, her status constantly monitored by my medical assistant.

But because of the Stark Law, there was no x-ray unit in our building. Without the x-ray unit, I saw the patient and called the hospital to schedule an x-ray. With an ever increasing frequency, the hospital radiology department informed me that I could send the patient to them but they had no idea how long the patient would have to wait before being seen. This wait was sometimes in the order of hours. That's a long time for an 86 year old patient in pain.

Then I had to find a means of transporting the patient to the hospital. At best, if she was with a son or daughter, a younger adult could drive her and stay with her. At worst, if she was with an elderly spouse or family member, my medical assistant helped load the patient into their vehicle and off they went to the hospital, hoping that there would be someone to help them unload. That was not always the case. In either circumstance, the transfer to and from their vehicle.

Once the x-ray was completed, the entire process was

reversed and the patient returned to my office. Because of the wait in radiology, this was sometimes after hours. Then, I had to try to get an x-ray report. The other alternative was for the patient to bring a *copy* of the x-ray with her. Hospitals no longer allow original x-rays out of the radiology unit, only copies. A copy, however, involves a fee, often not covered by insurance. It can be remarkably hefty. So, most times I was then required to call radiology and ask for a report because the patient had refused to pay for a copy. Sometimes the film had not yet been read. Then asking became begging, a plaintive pleading for whoever answered the phone to find a radiologist who would look at the film. If the patient's return was late in the day, the radiologist who had been reading "plain films" had often already gone home. Then, begging became swearing.

Finally, I was able to treat my 86-year-old patient, who by that time was exhausted, always in worse pain, and usually angry with me. I was always tempted to tell such a patient that their anger would be best served by writing Representative Stark a letter. I never did because both the patient and I were preoccupied with taking care of his or her problem.

This process was inordinately more complicated, took longer, cost more, and put the patient through more unnecessary pain than if there had been an x-ray unit in our building. Representative Stark's law caused many of my patients untold amounts of suffering.

The alternative was to simply send the patient to the

emergency room. Over the years, I turned more frequently to this alternative because I loathed having to expose my patients to a painful and exhausting marathon. This turned what would have been an encounter with me in my office that at most would have cost a hundred dollars to one that always ran into the thousands. As emergency rooms became increasingly overcrowded, it too became a marathon.

With regard to EHR, the Stark Law looks upon physicians' access to systems created by hospitals or insurance companies, without paying what would be considered a standard commercial rate, as unwarranted gain. For example, consider a hospital that wants to initiate the adoption of a universal EHR system. That system must include physician office records to be complete. That means that the physicians on the hospital's staff must adopt the use of the hospital's EHR system in their offices. However, even though universal physician participation is needed for a hospital to be able to implement a universal system of EHR, officials have threatened legal action if all physicians are not charged a commercial rate for the use of these EHRs. The rate is too steep for many doctors. Further, there is often debate as to what compromises a true commercial rate. Thus, even on this smaller scale of but one hospital, EHR systems remain undeveloped. There is talk about modifying the Stark Law to allow EHR, but at the present time, no action beyond talk has transpired. In the face of legal threats, many hospitals and insurance companies have abandoned any plans for comprehensive EHR.

There *are* some multispecialty clinics, large primary care clinics, and hospital systems that have EHR in place. The problem is that most of these systems have different formats, and can't exchange information. This means that any centralization of data—like lab results—is also impossible. It's as though every medical office, laboratory, emergency room, and hospital spoke a different language. The only way they can communicate with each other is through a translator.

It would seem that an obvious solution to this latter problem would be the choice of a universal format, but such an idea is anathema to free market philosophy. None of the clinics or hospitals that have invested tens or hundreds of thousands of dollars on their present EHR system is going to repeat the process for a universal format. Further, even the suggestion of *designating* a universal format has software companies cultivating relationships at the FTC. Can you say antitrust?

If all this weren't enough, to theoretically protect patients' privacy, the federal government has passed another law called HIPAA. This law established incredibly complicated rules regarding the transfer of information. It is the legislative equivalent of hunting quail with a bazooka. In its initial form, it prohibited a doctor from even calling patients by their names in the presence of other patients. Upon entering a doctor's office, a patient was to be assigned a number and referred to by that number—"Mrs. Number 219836, the doctor will see you now for your problems with your numbers 84 and 93 and the terrible effects these are having on your ability to do number 103."

Fortunately, someone in the *Government Office of Politically Correct But Totally Stupid Legislation* had at least as much common sense as a piece of liverwurst, and this part of the HIPAA Act was expunged. Unfortunately, what's left still makes any simple transfer of information difficult if not impossible.

But in the face of these problems, the President of the United States has claimed that the adoption of a system of universal electronic medical records is health care's panacea— even though this system will cost additional billions of dollars, even though the Stark Law prohibits including physicians' offices in systems funded by hospitals and insurance companies, even though HIPAA essentially prohibits these components from sharing information.

There is an *Alice in Wonderland* quality to the explanations of how these problems will be solved and how simply changing the structure of medical records will have such far-reaching effects. These explanations sometimes invoke magical thinking—"It will just happen." They may also be expositions of doubletalk—"The market forces inherent in the application of IT innovation may take incremental time units to engender a functional solution, a scenario with which we are all familiar, but that should not decrease the estimated chances of a positive outcome." Most commonly, however, they are a dismissive brush-off—"You just don't get it, do you?" Real solutions are rarely discussed because they do not exist.

The President need not look very far to find an example of what happens when the magic of modern data management develops a flaccid wand. The FBI advertised its new system of computers as "the most modern bad guy catching system in existence." The Bureau spent in excess of 170 million dollars, but at present, its agents can't even exchange e-mail. The entire system will need to be scrapped, and it's estimated that another 200 million dollars will be needed for a newer, new system—as it were.

The FBI is not nearly as diverse as the health care system. It is also a much smaller proposition than every hospital, doctor's office, and clinical laboratory in the entire United States. Extrapolating the failure of the FBI's computer system to the health care system makes me slightly nauseated.

I do not wish to imply that universal EHR is not important. While its development should be placed in its proper perspective, and the difficulties inherent in its implementation must be honestly evaluated, at some point it must become part of American health care.

Medical histories

One of my best friends is a general internist. He's the best general internist I've ever met. Not long ago, he told me about a patient of his. This fellow had gone to the emergency room. There, he had been swallowed by a system.

In the emergency room, he had been evaluated for chest pain. His evaluation had consisted of an EKG, cardiac enzymes (blood tests that show if heart muscle has been damaged), a chest x-ray, a helical CT scan of the chest (to look for a blood clot in the lungs), and the usual general lab tests. These were all negative. The patient's pain improved somewhat, but by no means did it go away. After five hours in the emergency room, he was admitted to the hospital because of chest pain of unknown cause.

All of this went on without my friend's knowledge. His partner had been on call, but his partner had only been told that the patient was in the emergency room with chest pain. The next message from the emergency room was that the patient was going to be admitted to the hospital.

My friend saw the patient early the morning after his admission. He talked to him for about two minutes. Even though he did not have access to all the results of the tests run in the emergency room, he immediately made the correct diagnosis. He ordered a confirming test and contacted a general surgeon to see the patient.

What could my friend have possibly done in two minutes that had not been done in five hours the prior evening?

He did an *accurate history*. He discovered that the patient had not had chest pain at all! The pain was actually in the area of the liver. The patient said that he had described his pain as "kind of like chest pain but not really." Further, the pain had started

161

ninety minutes after the patient had allowed himself a rare treat—a pizza. This is a classic history for an attack of biliary colic (gallstones). My friend did a physical test I call the "gotcha finger." The index finger was pointed at the patient's abdomen and then poked in the area under the right ribs. The patient responded by telling my internist friend he had just reproduced the pain and to stop under threat of death.

My friend ordered an ultrasound that confirmed gallstones. The gallbladder was removed that evening through a laparoscope, and the patient went home the next day.

The costs of the tests in the emergency room ended up being in excess of eight thousand dollars. I'm not sure what my friend charged for taking an accurate history and poking the patient's abdomen with one finger, but I am absolutely sure it was much, much less than eight thousand dollars.

As I said, my friend is a very good general internist. However, he did not tell me this story to brag about his diagnostic skills. He told it as an example of how little attention is paid to patients' histories. The emergency room physician thought he heard the patient describe chest pain and immediately initiated a workup for a possible heart attack. He initiated a *system* designed to evaluate chest pain.

Granted, chest pain is a symptom that may indicate life threatening problems with a person's heart or lungs, but this particular patient would have given a classic history for gallbladder disease—had someone listened to him. Fortunately, in this case the patient was not harmed. But eight thousand dollars of needless tests were performed.

Obtaining an accurate history is still a skill that must be learned and practiced. There is no technological substitute for a good history. It is ironic that many health care experts want to spend billions of dollars converting to electronic medical records while ignoring the importance of accurate histories.

A medical history is a description of symptoms, associated activities, past events, family history, and work history that is the first piece of data obtained by almost every health professional who has contact with a patient. The emergency room admitting clerk asks, "Why are you here?" The medical assistant or office nurse asks, "What's the doctor seeing you for today?" The doctor asks a similar question and then follows up with a series of other questions, trying to elucidate the nature of a patient's problems or how a patient is doing with relation to treatment.

Fortunately, this is not like questioning suspects in crime investigations because patients are not trying to hide anything. It is obviously in their best interests to relate an accurate history so that they are treated correctly. Thus the key to obtaining a good history is simply asking enough questions to obtain the necessary data. Right?

Sorry. Wrong. Efficiency experts would like to think obtaining a history is a straightforward collection of data.

Clinicians who think likewise are in deep trouble—and so are their patients.

Patients are human beings, and human beings have emotions, and emotions are the wild card of human behavior. They are the building blocks of human nature. Just as my worrying about my daughter trumped my identity as a doctor, patients' emotions and human nature can trump their ability to relate an accurate history.

For example, consider our patient Dan and his chest pain. He was the same age as his father when his father died of a heart attack. The death of a parent is a powerful dynamic that can affect people for years. Dates associated with such an event can trigger intense emotions often without the patient being aware of their origin. These emotions may be manifested as physical symptoms. This is called an "anniversary reaction." At the same time, chest pain itself is a powerful dynamic. It is a symptom that has many benign causes, but society advertises its potential seriousness in an aggressive and pervasive manner. I have had patients as young as seven who were convinced they were having a heart attack.

So Dan was swimming in a maelstrom of emotions and symptoms by the time the doctor entered the exam room. Because of circumstances, Doctor Jones did not have the opportunity to sift through the confusion, and the subsequent consequences were disastrous.

Emotion is not the only element of human nature that makes obtaining an accurate medical history difficult. Patients describe symptoms in a language specific to their own culture, experiences, and preconceptions. As we saw in the example of my friend's patient with "chest pain," a doctor must spend time translating what a patient says into what a patient means. Without this effort, symptoms can be dangerously misleading. I have had patients with chest pain whose history was easily defined in five minutes. I've had others who demanded thirty or forty minutes such that I could translate their histories into something I accurately understood.

Obtaining a medical history is a process shared by two people. One of these people may be in pain, frightened, confused, or angry. The other may be tired, overwhelmed, and also angry. Unless this reality of human nature is acknowledged, it makes little difference whether a medical history is stored in a sophisticated system of electronic medical records or written on an index card. Unless time, energy, and an understanding of human nature are invested in the process of obtaining an *accurate* history, it is at best worthless and at worst dangerous.

Medical Metamorphosis

As scarce as truth is, the supply has always been in excess of the demand.

-- Author unknown

CHAPTER ELEVEN

THE MOST ILL-FITTING SUIT OF ALL

Medical consumerism

As we have seen, the existence of an educated consumer who makes rational decisions is crucial to a system of supply and demand. It is the rational choices made by consumers that drive up quality and drive down cost. Thus, it is easy to understand why those who promote incremental change have rallied around the concept of medical consumerism.

The truth is: patients don't behave like rational consumers. Medical consumerism says: dammit, forget human nature and start shopping for health care the same way you shop for a new refrigerator!

The entire concept of medical consumerism and educated patient consumers is a bit slippery. Insurance companies are now publishing physician fees for various services and procedures. As we said when discussing HSAs, the experts suggest that cost shifting a significant portion of medical fees to patients will make them use this data to choose the cheapest provider—they will be "educated."

However, the pricing of medical services is already subject to the most severe controls of any professional service in America. All you have to do to substantiate this is look at any hospital bill. I recently had a former patient show me his bill from surgery for prostate cancer. The bill came to $34,000. His insurance paid $3,400. He was flabbergasted. "My God, Doc," he said with his eyebrows raised to the back of his head. "That's ten cents on the dollar." This type of discounting, apparently to an astounding rate of 90%, is all part of the game.

The same discounting is true for physicians. In the past few years, if I was asked how much I charged for an office call, I honestly answered, "I'm not really sure." What we charged was irrelevant because the Denver area is heavily penetrated by managed care. Companies like Pacificare and Anthem-WellPoint now simply tell a doctor what they are going to pay—take it or leave it. The idea that a patient can use a list of posted charges to do comparison-shopping is ludicrous. The charges are in reality *fixed* by the organizations posting them. I've no idea what prices the insurance company has decided to release to the public, but I suspect they are the highest they can find rather than what the physicians are actually paid.

It's a bit like the old farmer who boasted he had two pigs worth a quarter of a million dollars each. When people scoffed, he referred them to his neighbor for verification. The neighbor substantiated the claim, explaining the pigs had been his, but he had traded them to the farmer for a horse worth a half million dollars, thus establishing bragging rights for both farmers. (In today's world, these sorts of bragging rights are not available because the IRS would have both farmers pulled in for an audit the first time they puffed out their chests.)

The closed system between the two farmers created their bragging rights, but it hardly established a real market value for the horse and pigs. No horse trainer would write a check for a half million dollars to buy the horse, and Hormel is not likely to spend a quarter of a million on each pig. The same problem exists with medical consumerism. It is absolutely misleading for an insurance company to publish fees so that patients will have the "appropriate information" to participate in a system of consumerism when the insurance companies are the "market force" establishing the fees. They are like the farmers establishing the price of the pigs and the horse—it's a closed system.

It's also a system controlled by forces other than physicians. Insurance companies and the government establish physician's fees, not physicians. There may have been a time when doctors had bargaining power, but that time has passed. In truth, physicians' fees are the only ones that are controlled in all of health care.

The cynics among you are probably saying, "Baloney. I read all the time that the rise in health care costs was driven by an

169

increase in provider costs." My response to all you cynics is that you are not cynical enough, because you believed America's greatest hypocrites—the experts. What increases is the *total cost of services*, not the fees themselves. The total cost of services goes up because more of them are provided. More of them are provided because, primarily, our population is aging. The older the population, the more medical care it needs. This sounds simple because it is. It's a reality imparted by Mother Nature— old bodies don't work as well as young bodies.

If America stopped making new cars, in a few years the cost of transportation might be increased by a rise in automotive repair costs. Would this be because auto mechanics were ripping off the public? Not if what they could charge was controlled the way physician charges are controlled. It would be because there were a lot more clunkers on the road, and old clunkers need to be fixed more often than new cars. I realize I am running the risk of comparing senior citizens, and that includes me, to old clunkers. I apologize. However, face it senior citizens. Even if we've worked to make ourselves into "classics" akin to 1964 Corvettes, we still require more maintenance than a new Toyota. We cost more to keep running, and that is why *total* provider costs have gone up. It ain't because the doctors are rolling in newfound wealth. They just have more engines that need rebuilding, more brakes that need relining, and more electrical systems that need rewiring. They're working more so we can "keep on truckin.'"

Medical consumerism as a means of controlling health

care costs might be called a fairy tale if it weren't so hypocritical. It's no fairy tale. It's a cleverly designed piece of propaganda. It's designed to deflect patients' attention from the true cause of America's health care crisis—a market-driven industry that creates huge corporate profits and has failed to control costs.

Insurance companies set physician fees. Insurance premiums are set by ... guess who...insurance companies. The fox is not just guarding the henhouse. He's living in it. The concept of an educated consumer would be valid if consumers, i.e. patients, could bring pressure to bear upon the most powerful forces that control the health care industry—insurance companies, for-profit hospitals, pharmaceutical companies, and the federal government—but they can't. The concept exists as a means of disguising the real nature of patients' only means of controlling health care costs. What is that means? Not seeking care.

What the experts are suggesting with their new brand of health care is that by shifting costs to patients there will be a tremendous pressure for patients to decide *on a financial basis* what care they need. Hogwash. What this forces patients to do is decide *if* they need care. This is simply a form of health care rationing, with those who can afford it receiving care and those who cannot afford it going without. The concept of consumerism is a ruse disguising this fact. It camouflages rationing on an economic basis, disguising it as patients empowering themselves to become "partners" in their care.

171

If a patient has diabetes and heart disease but can only afford medications for one of the two diseases, which disease is she supposed to choose? Give me a break! That is a totally different proposition than deciding between an expensive used car and an inexpensive one. Anytime a patient begins to try to make the diabetes or heart disease decision on a financial basis, they will suffer.

As will society because those sorts of decisions lead to avoidable complications, and avoidable complications lead to avoidable health care costs in excess of what it would have cost to avoid them in the first place.

Mother Nature loves to gloat by whispering, "Pay me now or pay me later—cheaper now, expensive later."

When insurance companies allowed physicians to choose a course of treatment for patients, physicians chose the best treatment. This was not always the cheapest treatment. Insurance companies did their best to twist physicians' arms by making them argue with insurance clerks, by dangling incentives in front of them, by penalizing them financially (carrot and stick), but the two halves of the doctor/patient relationship still would not jump through the hoops of a market-driven system the way some of the economists wanted them to.

So, after some thought, it was decided that removing the doctor from the system would change everything. Shift a much greater part of the cost to the patient, convince the patient that when educated they are just as qualified as doctors to decide what they need, but keep the ultimate responsibility for treatment decisions on the shoulders of the doctors, and voila, you have a system with a new source of revenue—patients' pocketbooks.

Further, any responsibility for individual medical failure, like a patient's demise, is shifted away from the system. If something horrible happens because patients can't afford a treatment, the system can say, "Well, that was their decision." If something horrible happens because patients choose a cheaper treatment, the system can say, "Well, that was the doctor's fault. He should have made it clear that in this case the more expensive treatment was the best choice. Sue him."

All of this is accomplished without a reduction in the profit margins of any of the big players—the insurance companies, the for-profit-hospitals, and the pharmaceuticals. The government has not been forced to face the realities implied by statistics identifying our health care system as second rate and the incredible sidebar that 45 million Americans don't have any health care coverage at all. And the academicians, economists, and social planners who earn a living by designing systems can remain employed because now they must evaluate the impact and efficiency of this new system as well as design other systems to educate patients so that they can feel educated. P. T. Barnum would be proud. However, when he reportedly said, "there's a sucker born every minute," he was quite conservative in his estimate.

An educated patient

I had a patient I will call Hazel. She was a seventy-year-old, retired fifth grade teacher with a propensity for treating everyone as though they were fifth graders. She was the organist at her church, and since the death of her husband and her retirement as a teacher, this was the primary passion in her life. She was also dedicated to medical consumerism.

Hazel had a Dupuytren's contracture in her right hand. This is a condition where tendons develop chronic inflammation and over time they shorten. In its most advanced form, a hand with Dupuytren's can become horribly deformed with fingers curled into useless claws. Even though her hand progressively worsened, Hazel elected to treat it conservatively, stretching her fingers three and four times per day. But eventually she was unable to play the organ. She finally made an appointment to see if surgery might help.

Surgery for Dupuytren's contractures fails more often than not. Hazel knew she would never have a normal hand, but said, "I need enough use of this silly hand to at least plunk out a melody." She had Medicare insurance with a supplemental policy from her school district. This meant I had my choice of hand surgeons, so I recommended her to the best technical surgeon I knew. My feeling was that he had the best chance of restoring Hazel's right hand to some degree of function.

When I told her whom I thought she should see, she nodded, and then tilted her head, looking at me over the tops of her glasses. I suddenly felt like a fifth grader. "Now Doctor Waggoner, I will not tolerate waiting as I've had to in some other doctors' offices. This man must have enough respect for his patients to be punctual. Also, I plan on comparing what he charges to one or two other hand surgeons whose names I got from a referral service."

I winced. My specialist was anything but punctual. A number of times per week he was called by emergency rooms because of catastrophic injuries. He was one of the most skilled microvascular surgeons in Colorado, and had gained notoriety for his success reattaching severed body parts. With regard to cost, I doubt he had any idea where his charges fell in relation to a competitive market. He was literally a genius in the operating room but paid little attention to the rest of medicine's trappings.

"I understand your concerns, Hazel," I said, trying to sound like a competent physician and not a twelve-year-old. "How important are those requirements?"

She harrumphed. "Well I wouldn't have mentioned them if I didn't think they were important."

I shrugged. "In that case, we might have a problem. This particular surgeon is perennially late. As a matter of fact, I was going to suggest that you take a book with you to his office appointment. He's always being called out to put hands back on or reattach a finger. As for price, I've no idea. He's so busy I doubt he pays much attention."

"Unacceptable," she said. "Unacceptable. I'll need

another name."

"Okay. I understand. I know of one or two others who will get you right in and participate in just about all insurances. They are well trained and board certified in hand surgery."

Still looking over her glasses, she said, "Fine. That's what I want. I'm sorry to be a fuss, Doctor Waggoner. Perhaps other patients can tolerate your friend's way of practicing, but I can't."

I thought for a moment, and then said, "He's not really a friend of mine, Hazel. I barely know him. I just know his work. So does the rest of the medical community. That's why his waiting room is always busy. There are more patients who need his skills than there are hours in a day. That's what happens to good doctors, Hazel. The world finds out about them and their lives cease to be their own." I gave her the names of two other surgeons, both local and both qualified. I knew she could get in to see them quickly.

The next time I saw Hazel was ten or twelve weeks later. She was in for a check on her hypertension. As soon as I was in the exam room, she raised her right hand and wiggled her fingers. My mouth dropped open. "Look at that! Good golly, you're a veritable nimble fingers."

I walked to her side and examined her hand. An incision still red from healing ran across its palm. As she flexed and extended her fingers, it was obvious that their range of motion was limited, but there was at least motion. I nodded in approval. "Look at that."

"Strangest thing, Doctor Waggoner," Hazel said. "I went to the young hand surgeon across the street and after examining me, he said he would be afraid to try anything. He sent me to the first doctor you referred me to. I've never heard of that, a surgeon sending a patient to another surgeon."

I tried not to look stunned, but I obviously failed because Hazel laughed. "I can see you're a bit surprised as well, Doctor."

I smiled and shrugged. "How did everything go? Did you take a book to your office appointments?"

"Yes I did, and I needed it. But didn't you get a letter? I told him you were my doctor."

I shrugged again. "Not yet. I will, but he's pretty slow with those as well."

"That man! He is something else." There was obvious affection in Hazel's voice. She wiggled her fingers. "But he certainly performed a miracle on my hand. I'll be eternally grateful for that. Now that I have the use of my hand, I realize I'd have been an absolute fool not to go to the best surgeon available."

Eventually, the doctor who had refused to operate on Hazel became my primary hand surgeon. He had the perspective to know his limitations and valued the welfare of a patient more than a surgical fee. This proved to be indicative of his excellence as a physician.

Hazel's case embodies the problems with the concept of an "informed consumer." I've no way of really knowing if either of the second two surgeons I referred her to could have also successfully operated on her hand. One of them, however, did not even think so himself. I appreciated Hazel's desire to see a doctor who charged a reasonable fee and did not make patients wait. Those were reasonable requests, but they were superceded by her particular circumstance. I had to optimize her chance for success because unsuccessful surgery would have made her condition much worse, and further, there would have been virtually no second chance. The situation demanded she see the *best* technical surgeon available. I have had other patients whose Dupuytren's contractures were not as severe or whose hands were not involved in activities as precise as playing the organ. I had no problems sending these patients to surgeons who might not have been as gifted technically.

No two patients are exactly the same, nor are any two doctors. Appropriately matching them often defies the maxims of consumerism. It also often determines the success or failure of medical treatment.

Lead, follow, or get out of the way.

-- Thomas Paine

CHAPTER TWELVE

ANSWERS

The *well*-fitted suit

In 2007, the state of Colorado created a blue-ribbon panel to consider possible solutions for Colorado's health care crisis. Following a monumental effort of sorting through suggestions from all sides of the issue, the panel synthesized four possible models that addressed the cost of care and Colorado's uninsured.

The response to this effort was rancorous debate. Letters to the editor typically took the form of, "How could you be so stupid as to have suggested plan A (or B, C, or D)? That plan has been attempted in Canada (or Great Britain, Bulgaria, California or Massachusetts) and it failed miserably. People there have to wait years for life-saving surgery (or pay enormous taxes, endure medieval forms of care, or swim the Detroit river to America where they can get good care.)"

If you were not supporting one of the plans the way a sports fan supports an NFL football team, i.e. if you were someone trying to actually solve the problem and not grind an axe, it was easy to come to the conclusion that *no* plan could possibly work because they had all been tried, and they had all failed. Since these plans were supposedly a synthesis of *all* possibilities, that was a *depressing* conclusion.

But there *have* been successes. There have been well-fitted suits that America could wear without tripping over cuffs, and they have been found in a most surprising place.

Until about 10 years ago, the Veteran's Administration Hospitals (VAH) system was viewed as a source of medical care to be used only if no other was available. 10 years ago, VAH began a climb that eventually placed them as the American hospital system ranked highest for quality of care .

Additionally, the average annual cost for the care of a VAH patient fell to $5,000. This is $1,300 less than the $6,300 per annum cost for an average American. This is almost miraculous because the population of VAH patients is older and sicker than the population of healthy children and adolescents, vegetarian marathoners, and various flavors of health fanatics that make up the general population. If the VAH's annual cost were projected to a general population, it might be as low as *$3,000 or $4,000.*

This represents what many consider an impossible accomplishment—a system that has dropped cost while increasing quality.

The system turned around when its leadership was taken over by Doctor Kenneth Kizer, an emergency room specialist. He was appointed as the VAH's undersecretary for health in 1994. His background also included a stint as a Navy diver and a health official in California. I consider both of these to be death defying occupations, so perhaps it was his prior work history that gave him the courage to take on the largest medical bureaucracy in America and turn it upside down.

Doctor Kizer's first decision was that every VAH patient had to have a primary care doctor. He then centralized the purchase of supplies, divided the system into territories, each with its own defined goals, and moved the emphasis of the system from one that was hospital based to one that was primary care based and centered on community health clinics.

He also looked at the VAH's system of electronic medical records. The VISTA system had been put into service in the 1970s, but it had stumbled along for decades. Doctor Kizer upgraded the hardware and gave a mandate to the VAH software specialists to streamline the system and add components capable of patient recall, standardizing protocols such as diabetic care and immunizations, and tracking lab results. Since the software programmers worked directly with the health care providers who used their programs, changes were functional and implemented with relative ease (when compared to the chaos of the general medical community).

From the beginning of his tenure, Doctor Kizer utilized the creativity of his staff. A long time VAH nurse suggested a bar code system to match medications to individual patients. Her inspiration for the system was watching car rental agencies use a similar system for a streamlined return of automobiles. This innovation significantly reduced the incidence of medication errors at the VA H and is now used in other hospitals all over the country.

The dramatic transformation of an organization as large as the VAH—it is the nation's largest hospital system—was possible because of leadership that recognized what needed to be done, had the courage to initiate significant change, and actively utilized the individual skills of employees. *No system is any better than the work of the people who must implement it.*

During Doctor Kizer's time as head of VAH, thousands of veterans turned to the VAH health care system instead of America's private system, *even when they had private insurance.* Why? Because the VAH system offered better care.

Congress responded in a strange fashion. Rather than touting the system's success, it limited new patients to those with service-related injuries. Legislators appeared worried about the success of one of their own systems, in spite of it being the holy grail of health care systems, one that can increase quality while decreasing cost.

But, it did not adhere to the tenets of an unfettered free market. Opponents complained that they didn't want the VAH system to become "too big." (an 800-pound gorilla sitting on the living room couch next to suitors from insurance companies, hospital corporations, and drug companies.)

You're probably reading this, shaking your head, and saying to yourself, "Wait a minute. What about all that stuff with Walter Reed Hospital and the horrible things that happened to those kids just back from Afghanistan and Iraq? If the VAH system is so good, then how come they were investigated by everybody except the Department of Agriculture?"

In 1999, Doctor Kizer was forced out of the VAH system. He turned in his resignation after his reappointment was held up by Senator John Kerry of Massachusetts. What was Senator Kerry's beef with doctor Kizer? Would you believe money?

When the entire system was reorganized to increase efficiency, Massachusetts suffered a significant loss of revenue. This did not sit well with Senator Kerry. To quote him in a letter written to President Clinton, "My concern has always been helping our veterans find the health care they need. The Undersecretary for Health must be an advocate for the health care needs of all of our veterans. It is our shared responsibility to find a lasting solution to the VA health care funding crisis. I remain committed to that goal."

Let me pose one question. Would Senator Kerry have written a letter challenging Doctor Kizer's reappointment if the cuts in hospital funding had been in hospitals in Missouri or Colorado?

I think not.

Doctor Kizer was also unpopular with those entrenched within the VAH. Little wonder. He did everything short of

183

putting sticks of dynamite in people's pants. But Willie would have died had his leg not been amputated, and the VAH was considered an anachronism before Ken Kizer altered its way of doing business. There was talk of eliminating the VAH entirely. Like Willie, it was near death. There are circumstances that require dramatic surgery.

Kizer has been gone from the VAH for eight years. During that time, while some of the changes he instituted have remained in place, the forward-looking attitude that "forced an elephant to do a pirouette," as described by Maureen Glabman in *Managed Care Magazine,* is no longer quite so forward. Returning veterans from the Afghanistan and Iraq wars have faced the same bureaucratic inefficiencies and in some cases abhorrent physical facilities that characterized all of the VAH before Doctor Kizer transformed the system. Politicization derailed change.

However, an article in southofboston.com by Maureen Boyle sheds light on how difficult it is to manage any health care system. As controversy raged about follow-up care in the immediate vicinity of Walter Reed, she wrote an article titled "Despite scandal, VA care lauded by vets." It documents the positive attitude of veterans about VAH care in the areas adjacent to Boston.

To quote one veteran, "The VA, in general, has done just about everything in the world for me..." Obviously, there is adequate funding for the VA hospitals in Senator Kerry's state.

It is impossible to say, but one must wonder if this would have been true had Senator Kerry not changed the funding priorities back in 1999. He certainly had the power to do so. There was a Democratic president at the time, Senator Kerry would go on to become the Democratic nominee for President five years later, and the senior senator from Massachusetts was Ted Kennedy, a powerful figure in his own right.

So, Massachusetts received its local funding, but the change initiated by Doctor Kizer ceased.

It is also impossible to say whether Doctor Kizer could have averted what has become a crisis had he remained the VAH undersecretary. I personally think he would have. Doctor Kizer's regime at the VAH was the most progressive and anticipatory in the organization's history.

VAH—a well-fitted suit for everyone?

Would a system like Doctor Kizer's VAH work for all of America? Probably not. It is a closed health maintenance organization (HMO). A closed HMO is a health system where all patients are part of one large system. That system is all inclusive of hospitals, doctors, and medical services. The level of services are controlled centrally as are funding and protocols.

The Kaiser Permanente system is a closed HMO. In truth, so is the NHS. What is the NHS? The NHS is ...scary organ music, please ...Great Britain's National Health Service.

This is the point at which free market mavens,

representatives of organized medicine, most political candidates, and the vast majority of Americans whip out either a silver cross or a large piece of garlic to ward off the monsters known as ...once again, scary organ music, please, ...*socialized medicine and single party payer,* the Dracula and Frankenstein of health care plans.

Ken Kizer proved that a closed HMO can work. He also proved that a closed HMO can't work. "Aha!" you say. "Dumbo has finally proved something himself—that he is an idiot! How can something both prove and disprove something else?"

I suppose it really can't, but what Kizer did was demonstrate the problems inherent to a health care system where the supply side is controlled. Those problems stem from human nature, in particular the human nature of medical professionals.

For 18 years, I was the managing general partner of a general partnership of 32 physicians that owned a medical building. I took the job basically because no one else wanted it. They were all smarter than I was. It was one of the worst experiences of my life.

Trying to organize physicians and get them to agree on anything is akin to herding cats— *feral cats.* Medical professionals are independent, stubborn, and often tired and grouchy. That's good for patients because it is those characteristics that allow doctors to battle the unrelenting foes of death and disease. It's bad for anyone trying to manage them. Good doctors do not cotton to being told what to do.

It's also true that being a good doctor requires working very hard. I'm not talking about being a *great* doctor—just a *good* one. A system that does not reward hard work promotes not working hard—socialism's the basic flaw. Most people will not bust their backsides if the chap next to them is sitting with his feet propped up. That's human nature.

Together, these reasons explain why a system like that of the VAH was doomed to fail. Even if Ken Kizer had continued as VAH's undersecretary for health, at some point his ability to motivate the VAH employees, particularly the doctors, would have waned. He would have run out of sticks of dynamite to place in peoples' pants or someone would have placed a very large one in his.

This is why the closed HMOs of the world are trying to inject competition into their supply side. Both Canada and Great Britain have realized that their health care systems are sluggish and often inefficient. Both countries are trying to privatize parts of their health care delivery to create competition, with rewards for better performance.

So perhaps the garlic and silver crosses are appropriate. Whatever you want to call a closed HMO, it will never function to produce the highest quality health care.

Directed Demand Health Care and a Requisite Insurance Source—DDHC and RIS

Hopefully you are convinced that incremental change is

no change at all, even though there are a great many powerful people spending a great deal of money trying to convince you otherwise. When you encounter those efforts, trust your common sense. If what you're hearing sounds like someone describing plans for a jetliner constructed from concrete—don't climb aboard.

I would hope that I have also convinced you that neither an unfettered free market nor a closed HMO (I prefer that term because it avoids the large pieces of garlic) will produce the highest quality health care at the lowest cost. The fact that America, Great Britain, and Canada are all struggling to change their systems is ample proof that both of these assertions are true.

What is needed is a health care system that allows a free market to behave like a free market. The economists understand that this would be possible if patients behaved like rational consumers. They have tried to bend patients into those rational consumers. The economists have yet to accept the fact that their bending will never work.

I find it ironic that when the subprime credit crisis developed, the economists wailed about how irrational investors became when "fear was in the air." I would suggest that the fear of losing part of one's retirement is miniscule when compared to the fear of losing one's life. PATIENTS ARE IRRATIONAL. THEY ALWAYS WILL BE BECAUSE BY DEFINITION THEY ARE SICK AND FEARFUL.

I apologize for stooping to a childish use of capital letters.

If you are still reading this book, you probably don't need me screaming at you. But there is no way of getting around this very simple truth, so why is it ignored?

However ...um, organist, please hit that bit from the *Halleluiah Chorus* ...there *is* a way of creating a rational consumer. There is a way of injecting economically sound reasoning into the demand side of health care's system of supply and demand.

I call the process that would accomplish this feat *Demand Directed Health Care (DDHC)*. I call the rational consumer a *Requisite Insurance Source (RIS.)*

Consider eliminating America's vast array of insurance companies and replacing them with a single entity. This entity would be a *RIS*. It would be financed by state and federal governments, employers, and patients. It would handle *all* of America's health insurance.

Those of you paying attention may be saying, "Now wait one minute! That's a single party payer." Let me reassure you it is not. The concept of a single party payer implies control of the supply side of health care as well as the demand side. DDHC leaves the supply side untouched.

We demonstrated that health care has none of the components necessary for an effective system of supply and demand. Consumers ain't consumers; vendors ain't vendors; and health care sure as hell ain't laundry soap. We have also described myriad ways that the experts have tried to make the ill-fitting suit look like it fits.

DDHC proposes that it is possible to generate an effective demand side for the health care system by creating a consumer that is educated, empowered, and objective. That is what a RIS would be. Semantically, a RIS is a single party payer. In truth, it is the ultimate educated consumer. It is what the mavens of free market philosophy have unsuccessfully tried to create by bending and twisting patients into something they could never become.

The idea that market changes can be engendered by isolated patients, acting with a consumer's power of one, confused by conflicting medical opinions, and making decisions when they are in their least rational and most vulnerable emotional state is just plain ...stupid—sorry, I can't think of a word that is more appropriate.

A RIS would have access to the nation's best medical experts, so it would certainly be educated. Because of its size, it would have the financial clout to control costs. It would be empowered. It would be nongovernmental and not-for profit, so it could be objective, free from the pressures of special interests and focused on its mandate—high-quality, inexpensive health care. In short, it *is* the ultimate consumer.

Think back to the quote from *The Economist,* "Neoclassical economics is built on the assumption that humans are rational beings who have a clear idea of their best interests and strive to extract maximum benefit ...from any situation." A RIS is the *only* way to introduce rationality into the demand side of a health care system.

Having run through this concept with hundreds of people, I know that right now, many of you are thinking to yourselves, "What the hell is he talking about?" I appear to be mixing apples, oranges, and mud. Let me start over.

Demand Directed Health Care (DDHC) is a concept that proposes eliminating all medical insurance companies, *including the federal government's Medicaid and Medicare programs,* and replacing them with a single entity called a Requisite Insurance source (RIS). A RIS would be funded by state and federal governments, employers, and patients.

A RIS would be structured such that it is:

- *Independent and stable.* It would be governed by a board of 8 members. Board members would serve an 8 year term. One member would be replaced each year. The new member would be nominated by the board and approved by the United States Senate. This length of term provides abundant and inherent stability in upper management but also an ongoing injection of "new blood."

- *Broad based and knowledgeable.* The board would be comprised of 2 experts in primary care, 2 members from the ranks of specialists, 2 members of the business or economics community, one medical ethicist, and one legal expert.

- *Nonprofit.* The only economic goal of a RIS would be maintenance of high-quality health care at the lowest possible cost.

- *Internally motivated.* Just because a RIS will not create profit does not mean that its employees and board members should not be well rewarded. This entity will handle massive cash flow and should draw the very best talent available, and that talent should be well rewarded.

Obviously, insurance companies would vehemently oppose this restructuring of the health care industry. For good reason—they would be eliminated. America's bastions of free market philosophy like the Cato Institute would view this as an assault on commerce.

However, as we pointed out, the "big players" in health care are oppositional. Any change that benefits patients will damage at least one of them. Limiting the damage to insurance companies is a logical choice because it will produce no collateral damage.

For-profit health insurance did not even exist until a little over a quarter century ago. It is the construct of an opportunity to create profit, where none existed, *without improving or adding anything*. Pacificare, WellPoint, and United Health care did nothing that had not been done in a less expensive fashion by the not-for-profit Blue Cross and Blue Shields that preceded them.

The attempt to convert Premera from its not-for-profit status is a classic example of a process aimed primarily at benefiting upper level management. There is no evidence that Premera's subscribers would gain from the conversion. There is no evidence that such a conversion *ever* increases quality. *None of the top 10 plans in the 2006 ranking of health plans by US News and World Report* were for-profit.

For patients and the rest of American business, nothing is lost if a RIS replaces these companies.

For those working at lower levels of the medical insurance industry, this restructuring may represent no more than a lateral shift from one company to another. A RIS will be created de novo and thus have an absolute need for those trained in the administration of health insurance. It will increase meritocracy because the implementation of such a large system will demand that ability be rewarded and competence acknowledged.

For those in upper level management with *functional* skills relative to health care administration, a RIS will also represent an opportunity. Those capable of making the system work would be appropriately rewarded, both financially and in job satisfaction as they play a role in creating what should be a model for health care delivery.

For investors and upper level management, a RIS will

quite frankly represent a disaster. Investors will need to find other havens for their funds. For those whose skills were exclusively related to mergers, leveraged buyouts, and high-level financial dealings, like the former CEO of United Health care and his $1.5 billion bonus, the golden goose will no longer be dropping eggs.

The elimination of an industry with a cash flow of $1.5 trillion would be unique in American business history, but so would its benefits. In just increased administrative efficiency, a RIS would reduce health care costs by as high as 30%. As you might expect, this number is vigorously debated by the insurance industry and free market advocates. They have spun numbers into a suggestion that there would be *no* advantage. One must remember, however, that these are the same forces that suggest when you break your ankle playing softball, rather than going to the nearest emergency room, you should call two or three and do comparison shopping. If the cheapest happens to be all the way across town, bite the bullet! You must be an informed consumer!

If one factors in no need to generate profit or return on investment to the economy of scale, a 30% reduction in administration costs is probably not an outrageous number. But, for the sake of fairness, drop it to 25%. If a RIS offers an immediate savings of 25% in health care costs by reducing the cost of administration and its related services, that represents an initial annual savings of *$450 billion.*

That ain't chump change. As pointed out by Woolhander and Himmelsten in the *New England Journal of Medicine*, that is more than enough money to provide health care for all of America's uninsured.[3]

However, DDHC represents a change that supersedes savings on administration. It allows health care to become an effective system of supply and demand. The supply side— doctors, hospitals, medical manufacturers, pharmaceutical companies, and all the other associated industries—would be absolutely free to compete. The demand side would be educated, objective, apolitical, and empowered, and these characteristics of the demand side would neutralize the distorted realities created by medical care being what it is.

Could a RIS run roughshod over the supply side? Of course it could. Would it? Never, because the mandate of a RIS is the highest quality health care at the lowest price, and to achieve that a healthy, competitive supply side is a necessity. A RIS could foster such a supply side because it would be free from the leverage of political agendas and short-term crises of contrivance.

Through a RIS, DDHC will also create a structure for implementing many of the changes suggested by economists, social planners, and academicians. Consider:

1. As we have discussed, because of the chaos created by a variety of varying formats, HIPAA laws, Stark laws, and cost, a national system of electronic medical records

---

[3]Woolhander, MD, MPH; Campbell, MHA; Terry, MD, and Himmelsten, MD; "The Cost to the Nation, the States, and the District of Columbia, with State-Specific Estimates of Potential Savings"; NEJM; Volume 349:768-775, August 21, 2003.

(EHR) will be slow in coming. But, *every* insurance claim would go to a RIS, and there would be instant de facto standardization to whatever format it uses. A RIS would thus represent the backbone of a national system of EHR.

2. Most quality assurance studies and a significant proportion of large disease studies rely on Medicare billing records (discharge diagnoses, etc.). A RIS would create a nationwide data bank not limited by the demographics inherent to a Medicare or Medicaid population. In other words, every medical encounter for every age group would be part of the RIS data base. This opens up avenues for medical research otherwise impossible.

3. Obviously, a RIS would be the proverbial 800-pound negotiating gorilla. It could sit across the table from pharmaceutical companies, for example, and truly affect the cost of drugs.

   a. But cost would not be the only factor a RIS could influence. The pharmaceutical companies *should* direct their research into vital areas that might not be as profitable as "me too" medications in classes of drugs that are big revenue producers, i.e. those that must be taken for a lifetime like blood pressure and diabetic medications. Vaccines and antibiotics may be one-time propositions, but aside from climate change, infection is still humanity's greatest risk. Think AIDS or bird flu or EBOLA.

b. By upping the reimbursement for new vaccines and antibiotics, a RIS could *direct* the supply side without needing some form of government mandate. The demand side (the RIS) will have said, "We are willing to pay a premium for these items because they may save millions of lives." This would be a decision based upon an empiric assessment of what is best for an entire population. In health care, such decisions are necessary, but now they are usually made too late to be effective without a frantic and expensive rush to judgment. A RIS could initiate preemptive free market competition for the development and production of crucial pharmaceuticals (or anything else medical).

c. In a sense, this is akin to what the Department of Defense (DOD) does for weapon systems, aircraft, etc. The difference is that the DOD, by constitutional design, is subject to a considerable degree of political control. As a part of the federal government, it is certainly sensitive to political pressure. For example, the attempt to achieve increased efficiency by closing military bases was hacked apart, just as Ken Kizer had his efforts hacked apart, by regional political agendas. A RIS, being an independent entity that is by design apolitical, would not be similarly influenced.

4. DDHC creates an environment where there is both the flexibility to look forward in a long-term anticipation of future needs and react to acute crises.

   a. In the aftermath of Katrina, there was very little private sector participation of the health care system. The recovery was primarily implemented through the efforts of various charities and elements of all levels of government.

   b. In a similar circumstance, a RIS could respond by raising reimbursement levels for the affected areas, thus creating an incentive for the private sector. Equilibrium would eventually be established between what a RIS paid and what the market demanded for the needed services. The private sector would use its resources to most effectively render those services, as it does in any setting, even though this one would have been created by a natural disaster.

   c. This would avoid the typical "dumping" of dollars through grants. It would be an incentive-driven supply side response to a unique market force created by a RIS. Further, since a RIS would only pay for those services rendered, it would avoid the inefficiencies inherent in government "guessing."

d. This is a means of addressing difficult problems by utilizing the creativity and resourcefulness of private enterprise.

Medical Metamorphosis

We have been educated to such a fine—or dull—point that we are incapable of enjoying something new, something different, until we are first told what it's all about. We don't trust our five senses; we rely on our critics and educators, all of whom are failures in the realm of creation. In short, the blind lead the blind. It's the democratic way.

--Henry Miller in "With Edgar Varèse in the Gobi Desert," The Air-Conditioned Nightmare (1945)

CHAPTER THIRTEEN

THE LOYAL OPPOSITION

The unspeakable horrors of ...

I sense hostility growing in some of you. I can hear mumblings, and ...wait I heard it, I'm sure I heard it. One of you said something about Canada, and right after that, someone else distinctly said something about Great Britain. Come on—admit it. And now ...wait ...what was that? Oh yes, don't deny it. I heard the other phrase. Yes I heard it. Someone finally uttered the phrase—SOCIALIZED MEDICINE!

As I mentioned earlier, that phrase reminds me of a 1930s horror movie. The peasants are all standing in the town square. It's dark. The only light comes from flickering torches. There is a general murmuring, and then someone says "Doctor

Frankenstein" or "Count Dracula," and all conversation ceases. The peasants look at each other with melodramatic terror and revulsion. Someone has uttered the dreaded name.

Doctors and politicians react the same way when someone says ...hold the organ music, let's not overdo it ... *socialized medicine.*

I've recently been playing the role of the American reporter in the horror movie—the person who doesn't believe in vampires, etc. I've been asking some of my medical friends, "Okay, so exactly what is socialized medicine?"

My friend reacts not unlike a peasant in the movie. He looks around, pulls out that piece of garlic on a string around his neck, and recites a medieval European prayer beseeching protection from evils of the night. Then he says, "Socialized medicine? Have you gone mad? Have you taken leave of your senses?"

I smile, like the reporter, and say, "Maybe. But really, what is socialized medicine?"

He rolls his eyes, look to the heavens and says, "Well, it's when ...I mean the government ...that's it ...the government tells doctors what to do. They tell doctors *how to practice medicine.*" Then he smiles and nods his head. "In places with socialized medicine, they tell doctors what they can and can't do."

I nod as well. "You mean like the insurance companies do here? Like when doctors have to call those clerks and beg to get an MRI precertified?"

His nodding stops. "No. It's different."

My own nodding continues. "How?"

"It's different," he says. "There, I mean, the thing is ...the government also tells you ...what you can charge!"

"You mean like Medicare? Like all the HMOs and PPOs already do here?"

By this time, other medical people surround us. I see one of them lighting a torch. "Independence!" one of them calls out. "Socialized medicine strangles our independence."

I take a couple of steps backward because I see a pitchfork being waved in the air. "You mean like it's already being strangled right now," I ask, "with almost no medical school graduates starting their own practices? With most going into huge groups or HMOs?"

Usually this is the point in the conversation when I'm hit with a piece of garlic.

America already has a health care system that would meet any of the definitions of socialized medicine that make doctors light torches, grab pitchforks and band together to roust out the monster. Loss of autonomy, rigid financial regulation, bureaucratic incompetence, and professional meddling are all day-to-day realities for every doctor in America. Dracula has already left his hicky on our necks. We just don't want to admit it.

But DDHC is nothing like the health care systems in Canada or Great Britain. Their systems have problems that would

203

persist even if financed to America's level of funding, the difficulties inherent to systems where hard work is not rewarded. These systems are inefficient because there is no incentive to *not* be inefficient.

The ultimate socialist doctrine is, "From each according to his ability, to each according to his need." Left to our own devices, most people will eventually react to that doctrine by deciding, "Great. I'm disabled—no ability whatsoever—and I need a sports car, a 98-inch plasma television, and that real cool beach house I saw in a magazine the other day. I'll drive the sports car to the beach. Deliver the TV once I'm there. I'll call. Oh, right. I also need a new cell phone."

Most human beings are hardwired to expect rewards for hard work, ingenuity, or skill, or whatever allows them to contribute more to society than the slug sitting on his front porch sipping a beer and eating chips for 15 hours a day before sleeping the other 9. If that reward never comes, productive citizens have one of two alternatives. They can move their couch out onto their own front porches and say, "Aw, the hell with it." Or they can set the slug's couch on fire, perhaps with him still seated on it. In the long run, neither possibility is healthy for society.

That's why DDHC leaves the supply side untouched. It allows the doctors and hospitals and medical technology companies to beat each others' brains out in that friendly game called the free market. If not regulated into paralysis, a competitive free market will find the most efficient means of solving problems.

The problem occurs when free market philosophy slips over the edge from being an economic theory to become a religion. In turn, this problem becomes a nightmare when that religion becomes fanatical. When that happens, any critical thought is impossible.

And that is where America stands with regard to health care policy. Critical thought be damned. You're either a capitalist or a communist. Either you're carrying a pitchfork and a torch or you're a minion of the devil.

And Americans die by the hundreds of thousands. And our economy is slowly garroted by a rope fashioned from health care costs. And nothing happens. The peasants die from a plague and America clings to a medieval theory called incremental change. How silly. How tragic.

An arithmetic lesson

A perfect example of how critical thinking about health care has been replaced by fanatical dogma is the comparison of Canada or Great Britain to America. Economic zealots who make these comparisons are in error because of their inability to do arithmetic. It always amazes me how experts can cite complicated statistical analyses supporting an illogical conclusion and ignore arithmetic that suggests a logical one—good, basic arithmetic.

Data reported on Nationmaster.com, summarized from the

major international data sources, reveals that in 2002, total health care spending in Canada was $2,931(international dollars) per patient a year. In the United States, per capita expenditure was $5,274. The United States spent 96% more per capita on health care than did Canada.

I must remind you—this is not rocket science. It is arithmetic.

America spends almost *twice* what Canada does on health care. If America adopted the Canadian single payer system and kept per capita expenditures where they are now, American health care would be a Canadian system with *twice* the resources. If Canadians spent as much on health care as America does, they could double the resources devoted to their health care system. Any comparison between the two systems must consider not only their structural differences, but also the amount of resources devoted to the systems.

How long a wait would Canadians have for specialty appointments (one of the classic criticisms of the system) if there were twice as many Canadian specialists? All things being the same, if twice the dollars were spent on each Canadian patient, there could be twice as many specialists. The same argument could be made for MRI units or any medical service or technology.

The same situation holds true for Great Britain except America spends *three* times as much as is spent in Great Britain. If one wishes to compare Great Britain's health care system purely on the basis of structure, since that system covers *all* Great Britain's citizens, since the measurements of longevity and overall health rank Great Britain above the United States, and since this is accomplished at *one third* the cost, then it is easily possible to come to the conclusion that the *structure* of their system is vastly *superior* to ours.

To criticize Great Britain or Canada because they don't have the medical facilities or the specialists that America has is absurd. They *shouldn't* have an equivalent amount of medical resources because they spend ½ **or 1/3** the dollars to fund them. If you spent 3 times as much on widgets as your neighbor did, wouldn't you expect to get 3 times as many widgets?

Citing differences in the number of specialists or x-ray units or any medical resource doesn't shed any light on why Great Britain and Canada are trying to change their health care systems. The flaw in the structure of Great Britain and Canada's health care system is in their *inefficiencies*. A socialistic supply side is always inefficient. History has ample proof of that fact, but comparing present day health care systems in America to Canada or Great Britain is not one of them. The disparities of funding are too great for any valid *quantitative* comparison.

American business leaders decry the cost of American labor. They constantly lament their competitive disadvantage in the global marketplace. Why then do these same business leaders refuse to even consider the implementation of some system that could dramatically reduce the cost of health care? It makes no sense. Those who claim to be advocates of a free market prop up the present system that threatens the ability of America to compete in that free market. This is irrational.

Had I practiced medicine in a similar fashion, my waiting room would have been littered with corpses. I most assuredly would have set some sort of record for malpractice suits.

America's only natural resource

In the period immediately following the cold war, Russia faced an economic crisis. It was her natural resources that allowed Russia to recover from what appeared to be the start of a long slide into a devastating depression. Now, buoyed by an increase in the price of oil and natural gas, the heart of the former Soviet Union has not only recovered economically but also started to growl like the bear of old.

What would happen to the United States if, for whatever reasons, it faced the same sort of setback? Could the United States rely on its natural resources as a means of recovery?

The answer is no. America must import her oil, gas, and metals—all raw commodities other than coal. At one time, America may have been a nation rich in natural resources, but no longer.

America has one natural resource and one only. If America were to experience an economic reversal similar to what Russia experienced, the only resource it could rely upon is its people. The only marketable commodities that America possesses are the ingenuity, the work ethic, and the creativity of its people. Unfortunately, there is every reason to believe that this resource too is being squandered.

Certainly America's health care policy would indicate that those who determine that policy have little or no appreciation for the importance of this natural resource. Think about it. How well will people perform in their jobs, how creative can they be, how well can they learn new technologies if they have untreated illness or chronic disease, if they are worried about the health of their children, or if they are in physical pain? How much entrepreneurial initiative is stifled by people staying in dead end jobs simply because those jobs provide health insurance?

And America's future? Without a healthy, strong generation of children, America has no future.

If an entire oil field in Alaska were set to the torch, do you think Congress or the energy companies would stand around debating theory? Hell no they wouldn't. They'd be doing everything in their power to put out the fires. Why then do Congress and health care experts stand around debating theory when 40 million Americans have no access to health care? Does anyone really believe these 40 million Americans are as productive as they would be if they *did* have health care?

We will never be a net exporter of oil. We will never be a net exporter of natural gas. Right now, we are not a net exporter of anything.

America was spared the horrific destruction wreaked on

the rest of the entire world during WWI and WWII. America did not have to rebuild its cities and industries. In fact, after WWII, America prospered because of the unbelievable boost the war gave to its businesses and industries. We are still coasting along on the momentum of that boost, but the car is out of gas—it's slowly rolling to a stop.

We need Americans to be productive. We need Americans to be creative. We need a generation of healthy, well educated children to grow into a generation of world-beaters. Right now we have none of this. The natural resource called the American people is being cared for with less attention than that directed towards our bridges and highways.

Democracy is two wolves and a lamb voting on what to have for lunch.

Liberty is a well-armed lamb contesting the vote!

  -- Benjamin Franklin.

CHAPTER FOURTEEN

DRAMATIC CHANGE

Unfettered ≠ Efficient

A completely unfettered free market health care system will chew up Americans and spit them out. It already is—at a rate of 100,000 premature deaths per year. But America's system isn't even unfettered. It is hobbled in ways that *prevent* efficiency and high quality while doing nothing to protect Americans from bad doctors, bad medicines, and charlatans. I would estimate that 15% of the forms I filled out over 30 years of practice served any function. The rest existed for the purpose of either covering someone's backside or justifying someone's unnecessary job.

That is, quite simply, stupid. There it is again, that word—stupid.

Demand Directed Health Care (DDHC) and the implementation of a Requisite Insurance Source (RIS) would free America from the inefficiencies that have accumulated over decades. A RIS would essentially be a *business* whose function was to act as the ultimate consumer. Even though there are those

211

who will characterize it as ...wake up the organist will you and cue him on the scary music ... *socialized medicine*, it is exactly the opposite. It will allow health care to operate as an effective part of a system of supply and demand.

Has anyone ever created an entity that can efficiently handle $1.5 trillion of cash flow? Probably not. But has there ever been a monopoly with $1.5 trillion of cash flow whose mandate was not to generate profit but to deliver the highest quality service at the lowest possible price while operating free of political pressure?

Never. It is hard to overestimate what such an entity could accomplish.

I am absolutely convinced that a RIS is the only effective means of injecting common sense into our health care system. I am absolutely convinced that it is the only effective change that will not create havoc.

Creating a RIS is simple. It involves impacting one single industry—health insurance companies. It is the only change that can be accomplished as a fait accompli. Compared to trying to dramatically alter the supply side of the system, it is a walk in the park.

The supply side must change, but its change should be one of an evolutionary nature. To put it a different way: the supply side should be changed incrementally. (And yes, I added that sentence for irony.) Once a RIS is in place, competition can do its magic towards the end of creating health delivery systems that meet all the wonderful criteria that everyone has been touting for decades.

I have no doubt that such an end will be achieved. Americans can be remarkably creative—but first they must be given the opportunity to do so.

$$$$$

The major eye-rolling question about DDHC, or any health care system, is, "Okay, Bozo. Who pays for all of this?" The answer is simple—the same sources that now pay for health care: employers, governments, and the public. The only difference conceptually is that the funding would all go to one source—a RIS—and be paid out by one source. In a structural sense, nothing else would change. Whatever other changes might evolve would be fostered by the interaction between the RIS and the supply side.

As I said, for two or three years, savings would need to be pumped back into the RIS to facilitate its implementation. Even though it is the simplest health care transformation, the creation of a RIS would be a mini-Manhattan project. Insurance companies could and probably will hire all but two or three of America's lawyers to file more suits than have been filed by Donald Trump's ex-wives. They will also employ *all* of the spin doctors to characterize a RIS's creation as a cross between a fascist plot, an act of economic terrorism, and an attack on motherhood. If the political yahoos ever get the courage to consider DDHC, it may take years to overcome the roadblocks, punctured tires, and blown bridges the legal yahoos create to protect the rights of the insurance mavens to collect their golden eggs.

But 400-plus billion dollars goes a long way towards solving problems. Once initial problems have been solved, the real savings can begin.

It would be crucial that DDHC not be seen as "free medical care." It may be best that a greater portion of patients' contribution to RIS come in the form of co-pays rather than premiums. Services that cost little or nothing are viewed as being worth little or nothing. This is also a means of addressing the present concern about the effect universal coverage will have on utilization of medical services.

With a schedule of co-pays designed to be significant enough to represent a real purchase but not so high as to inhibit appropriate care, patient utilization will reach an appropriate equilibrium. And with regard to physicians' over-utilization, a RIS would have the most potent weapon in existence to control patterns of inappropriate practice—data. With a single source of payment, the data base available to monitor physicians' modes of practice would allow checks and balances at every level.

This is why quality will also be enhanced with DDHC. With the data available through a RIS, it really *is* possible to monitor standards of care. There would be no more silliness of asking doctors to fill out data sheets to check to see if they are adhering to certain evidence-based criteria. If "x" diagnosis is supposed to be treated with "y" medication, a RIS will be able to check to see who treated patients accordingly. Those who did not can be asked why, and if their patients have done better, the criteria may even be altered. If their patients did not do better, then the doctors can be educated about why the criteria make sense.

This is a rational process, not a witch hunt. Is it Big Brother? In some sense it is. But I think most physicians would agree that if there is going to be an evaluation of their performances, it should be done by those seeking to improve patient care and not those trying to increase profit or pass the proverbial buck.

Ben's quote

I have opened each chapter with a quote. I did so because as I labored, I found the observations of wise people reassuring. If someone like Benjamin Franklin or Aristotle articulated a thought that was in concert with what I was saying, then perhaps I was not the idiot that I felt myself to be.

The quote from Benjamin Franklin is of particular relevance to the close of this part of the book. We are all well armed lambs because we still have the freedom to demand that common sense prevail over self-serving rhetoric.

Many lambs will need to contest against the wolves to

bring about the creation of a RIS. The wolves view a future without lamb chops as less than rosy.

STEP THREE

MAKE MEDICAL HUMANISM POSSIBLE

At present, medical professionals attempting to treat a patient as a human being must do so *in spite of* our health care system and not because of it. An emergency room clerk taking the time to acknowledge a patient's confusion, a nurse spending extra moments with a patient who is in pain, or a concerned family practitioner who does not adhere to the schedule of fifteen minute appointments scheduled every ten minutes will not be rewarded—they will be penalized.

Compassion, understanding, listening, relating, and empathizing are the core of the art of healing. They are also time intensive. In an environment where efficiency is infinitely more valued than healing, all of these are discouraged.

Further, over the last fifty years the profession of medicine has evolved into a *disease oriented* discipline rather than one that is *patient oriented.* This evolution has created a health care system that so fragmented that humanism is rare enough to be promoted as an exception rather than a rule— "Our doctors listen; our nurses care."

All doctors should listen. *All* doctors should care. Yes, doing so is difficult. It means one must open themselves to the suffering of others. But that is what defines the art of healing. It is a task that should be eased by a health care system, not impeded.

Without medical humanism, patients' suffering is greater than need be. Without medical humanism, patients' care is not as appropriate as it should be. Without medical humanism, America's health care system will always be chaotic and disruptive, regardless of its economic construct.

Fortunately, there are ways to make humanism possible. Not surprisingly, they are simple in nature and difficult to execute.

BONUS CHAPTER

If you have made it this far in this book, you have covered some difficult and disturbing material. Consequently, you deserve a reward. I am going to give you the reward in the form of a very short mystery—a discovery of information that in all likelihood your own doctor doesn't know. I say that because I certainly was unaware of it, as you shall see.

You may want to drop the information on your doctor the next time he or she asks you to put on one of those paper gowns designed to strip you of your dignity.

THE CADUCEUS CONSPIRACY

To be honest, I was never much of a *DaVinci Code* aficionado. I do know it's based on the premise that icons like the painting "The Last Supper" can have secret implications about profound issues. However, I lost interest when I found out that the book did not answer the only question I have about DaVinci's painting—why was everybody sitting on the same side of the table? I can't imagine someone not saying, "Hey, that's the third time you dropped one of your figs in my lap. Give me some elbow room. Move over to the other side of the table."

Now, I'm thinking about buying a copy of the book. I also watched an ad for the movie and might rent it. You see, I've stumbled headlong into my own DaVinci code. Actually, it has nothing to do with Leonardo. It concerns snakes and wings.

It began one day when my son came up to me, pointed his finger at my chest and asked, "What's that called?"

I looked down, paused for a moment and then answered, "Actually, it's called a tie. It's a men's accessory, worn on occasions that demand something more formal than a tee shirt..."

"Very funny," he said. "I meant the medical thingy on the tie. The thing with the wings and the snakes."

"Ah, you mean the caduceus." I placed my hand behind the tie, lifted it up, and regarded the upside down pattern of gold caduceuses embossed all over the tie. "It's the symbol of the medical profession. A patient gave me this tie."

"Caduceus," repeated my son. "Somebody asked me the other day, and I didn't know. What's with the snakes? And how come there are wings on the top of the stick...or whatever it is?"

"What?" I moved the tie closer to my eyes. "The snakes? The wings? They...they are symbolic of ...of myths...about...medicine." I smiled weakly.

He smiled back. "Just wondering."

When I took off the tie later that evening, I looked at it right-side up. The longer I looked at the emblem of my profession, the more bizarre it became. "Snakes?" I muttered to myself. "And what *is* with the wings? And how come there isn't even a sling or a bandage hanging off one of the wings or stuck on one of the snake's fangs?" A caduceus suddenly appeared to be a very strange item.

I realized that, basically, the only knowledge I could claim about the emblem of my profession was how to spell "caduceus." I had ties, cufflinks, paper weights, and golf shirts adorned with this concoction of reptiles and aeronautics, but I lacked the vaguest notion what it symbolized.

There was a time when musing is as far as my curiosity would have gone. Like vanishing single socks, there was an inexplicable phenomenon involving encyclopedia volumes— whatever volume I wanted would be missing. For example, if I had a sudden interest in George Washington, "W" would have vanished, reappearing only when I had lost interest in wooden dentures. I thus lived an existence stained by mismatched socks and an intellect starved for answers.

But no longer. Modern technology may not have solved the business with socks, but it has thwarted ignorance. Now, there is Google™. Elucidation is but a mere three or four keystrokes away, and the hardship of actually opening a book and recalling the alphabet while looking up a topic are no more. I hung up the tie, walked to my study, and googled™ "cadusus." Google replied, "Do you mean "caduceus?"

"I knew that!" I barked, perhaps a bit too loudly. "I just mistyped. I *know* how to spell 'caduceus.'" I answered yes and in .04 seconds, I was given access to 889,000 entries. I highlighted the first, one from Wikipedia, and read that a caduceus is "a winged staff with two snakes wrapped around it... an ancient astrological symbol of commerce and ...associated with the Greek god Hermes [the Roman god Mercury], the messenger for the gods, conductor of the dead and protector of merchants and thieves."

That puzzled me a bit. Merchants and thieves? And Hermes? I'm no great classicist, but I could not recall Hermes saving Athens from any plagues or trying to place Achilles in sandals with a heel lift for his tendonitis. I punched up another site and found a caduceus described as "Hermes' magic wand." The thought that the tie hanging innocently in my closet was covered with a repeating pattern of Hermes' magic wands seemed somehow debauched. The caduceus code began to have unsettling implications.

It occurred to me that perhaps after falling in with a bad crowd, Hermes had decided to clear his good name, attended medical school, set up something like the Mount Olympus Free Clinic for Indigent Gods, and that was why his magic wand had been adopted as the emblem of my profession. I googled™ Hermes.

There was lots of discussion of his "shrewdness and cunning," the fact that he had once been a phallic god, and how he was "fleet afoot," thus becoming the gods' designated messenger. There was nothing about medical care, compassion, or skills as a healer. It was quite apparent that my tie was covered with the symbol of a shrewd, cunning protector of merchants and thieves who also had enough foot speed to successfully escape the scene of any crimes committed by his thieving buddies.

I was stunned. The emblem of my profession, the medical profession, the profession dedicated to healing, committed to assuaging pain and suffering, was actually the magic wand of the god of merchants and thieves. I could not decide if I should laugh or cry.

When I began practicing medicine thirty years ago, money was a topic rarely discussed. The attitude was best summarized by one of my mentors—"Don't ever let your priorities get turned backwards. You can't serve two masters. Do a good job caring for your patients, and the money will take care of itself." Now, that attitude seems antiquated and naive. *Everything* is about money. The entire future of American health care is being plotted by economists and CEOs, not by physicians, nurses, or anyone who actually spends their days and nights caring for patients.

Of course, this transformation is denied, particularly by economists and CEOs. To discover that it is celebrated on my tie was ironic but also very discouraging. I winced at the thought of thousands of doctors cluelessly walking around carrying tiny billboards on ties, cufflinks, golf shirts, and pens extolling *commerce and theft*.

I googled a bit more and discovered that there was another reptilian symbol, the Staff of Asclepius. This one was not nearly as impressive as Hermes' magic wand. It was just a single snake wound around a piece of wood. There were no wings, and the wood of the "staff" was rough compared to the finished surface of the Hermetic wand. But *this* appeared to be the true emblem of the healing arts.

I discovered that Asclepius was a Greek physician who lived in approximately 1200 B.C. His skills were so widely celebrated that he was eventually deified. The staff's origin appeared to have come from treatment for *Dracunculus medinensis,* a filial worm. This parasite moves around a patient in superficial fat. To treat it, ancient physicians would make an incision just in front of the worm's path, and as the worm crawled out, wrap the worm around a stick. The infection was so widespread that eventually the image of a worm wrapped around a piece of wood became an advertising symbol for physicians. When the good doctor Asclepius was deified, the stick became a staff and the worm became a serpent, only fitting for a doctor who really *was* regarded as a god.

I also found out that only in America is the caduceus commonly used as medicine's symbol. Elsewhere, the Staff of Asclepius is considered the true emblem of the healing arts. That deepened my suspicions. America is the only industrialized nation without a national health care program, the world's only unfettered free-market health care system. It was also the only place that medicine's symbol was actually a designation of the power of commerce and theft.

Where was Tom Hanks when I needed him? The Caduceus Conspiracy? But before I could begin to worry about albino assassins (I saw one in the ad for the movie), I discovered the probable cause of America's adoption of the caduceus, and it was hardly a conspiracy. In 1906, the Medical Department of the United States Army mistakenly made the caduceus its official emblem. It had remained so since then, and apparently the rest of the country followed suit.

It was easy to imagine the head honcho army doctor saying to his aide, "Corporal, we were just funded for official stationary. I want you to get some printed. And look here, spruce it up with that symbol, the one that has a snake."

The corporal salutes, hurries off to research the symbol, and eventually comes across both the caduceus and the Staff of Asclepius. The caduceus is much sexier, and as tough as caduceus is to spell, Asclepius is tougher to pronounce. He chooses the caduceus and the rest is history. Hardly a conspiracy.

But the longer I thought about the whole thing, the more uncomfortable I became. So what if the Medical Department of the United States Army made an error a century ago? Why hadn't someone simply said, "Hey look, I didn't take call every other night for what seemed like an eternity so that I could wear cufflinks implying I'm a money-grubbing thief. Knock it off with the caduceus will you? One snake, one 'roughly hewn wooden staff,' Asclepius—he's our man. Forget Hermes, the wings, and the *pair* of snakes."

225

The answer to that question became apparent as I related my discovery to other doctors. Only one already knew what a caduceus really was. He also had a Masters in medical history. Surprisingly, he was also the only one who appreciated the awful irony of the circumstance. He smiled sadly and said, "Commerce and theft. Pretty telling isn't it?"

I've not worn my tie since the day I made my discovery. I've tried to rationalize that it's an expression of a patient's gratitude, but I can't. I've looked for a tie adorned with golden Staffs of Asclepius, but I've yet to find one. The fellow with the Masters in medical history told me they are available in Canada.

4

[4] http://commons.wikimedia.org

We have to ask ourselves whether medicine is to remain a humanitarian and respected profession or a new and depersonalized science in the service of prolonging life rather than diminishing human suffering.

--Elisabeth Kubler-Ross

CHAPTER FIFTEEN

HEALTH CARE'S EVOLUTION

Dogged Companions

If you are easily offended by tales of wild sex, uncontrolled drug use, under-the-table deals with pharmaceutical companies for hundreds of thousands of dollars, horrible mistakes covered up to protect reputations, vendettas against lawyers played out in the basements of hospitals, insurance executives forced to have embarrassing surgery and then blackmailed for full-color video tapes of the procedure, or how doctors have a drug that cures everything but keep the formula secret so they can stay in business, it's still okay to go on reading. As much as I would like to be able to write about those things and sell lots of books and go on talk shows and have a personal assistant named Suzanne whose only job is to keep me organized, on task, and looking good—I can't. Bummer.

What I would like to do instead is offer a brief overview

of medicine's history. When I say brief I mean the sort of brevity associated with a book report written about a book you didn't read. We're looking for functional concepts not historical detail. In other words, don't worry. I'm not going to drag this out to the point that you fall asleep. You may become drowsy, but I doubt that more than one or two of you will actually—you know—snore.

Pain and suffering have been part of the human condition since we first set foot upon the savannah. Our earliest ancestors led short, brutal lives, battling to live long enough to leave offspring—fighting to survive as a species. In part, it was our intellect that secured this survival. It was also our ability to form complicated social relationships, some of which centered on caring for injured or diseased members of our species. Skeletal remains from as long as twelve thousand years ago have revealed evidence of simple but effective care of the severe fractures and multiple traumas associated with hunting large, powerful animals.

As our intellect evolved, we began to experience pain different from the sensation of a broken limb or an infected tooth. We became aware of our own vulnerability and mortality. We began to appreciate the enormous magnitude of the universe and our inconsequential role within it. Haunted by these realizations, we created the disciplines of religion and mysticism, bastions against this new suffering of our spirit and soul. Physical pain and disease were understood within the context of these disciplines, often viewed as derangements of the spirit or as punishment for violating a belief system. Thus, ritual, ceremony, and the power of belief systems were added to humanity's earliest tools of healing. Even today, in tribal cultures a medicine man is both a healer and a religious figure.

Gradually our understanding of nature and the world around us increased, as did our efforts to directly affect physical suffering. For thousands of years these efforts were limited to the level of an empiric use of roots and herbs for their medicinal properties. Then, four hundred years ago, scientific knowledge exploded, and we peeled away the shrouds of superstition and ignorance. Roots and herbs were eventually replaced by antibiotics, immunizations, antidepressants, and heart medications. War, humankind's worst folly, may have continued unabated, but science discovered ways to limit that folly's consequences. The wounds of war begat anesthesia, plastic and trauma surgery.

Religion and mysticism accept as an eternal truth that pain and suffering will be our dogged companions. Modern medicine has at least partially domesticated the beasts, but in so doing has dismissed humanism as a healing force.

Without being biased, I believe it is possible to say that Americans are the most fortunate humans to have ever walked the face of this planet—free from the shackles of despotism, generally well fed, blessed with the freedom of at least some daily activities other than a constant struggle for food and warmth, and also free from most of the diseases that have historically ravaged humanity. American medicine plays a significant role in this good fortune. Its ability to manipulate the human body appears almost miraculous. Medicine has gone beyond curing and seeks to transform. We read of possibly controlling the process of aging. We have a complete picture of the key that literally defines us, the human genome. Medicine should be in a truly golden age.

Instead, it is described as being in crisis.

Disease-oriented health care

Prior to World War II, the war against disease and suffering was basically fought with swords and bows and arrows. Antibiotics were almost nonexistent. The pharmacological treatment of heart disease, hypertension, and diabetes was primitive. While surgery was often used as an aggressive mode of treatment for a variety of conditions, it often caused as many problems as it cured. WWII initiated a transformation of medical care and an exponential growth of knowledge and treatment options. When the general public began to reap benefits from wartime breakthroughs such as penicillin and new surgical techniques, a societal commitment to "modern medicine" was assured. As a result, disease after disease was conquered, and the "miracle of modern medicine" became an accepted euphemism.

America led this conquest of disease. In 1945, the United States stood as a technological giant. It unscathed by WWII. It had also reaped the benefits of the war's frenetic science. It had produced a bomb that by itself could level a city. Less than a decade later, it cured poliomyelitis, a disease that had paralyzed America's wartime leader, Franklin Roosevelt.

Technology irrevocably altered America's foreign policy. Ironically, it well may be that the development of nuclear weapons spared humanity another worldwide conflict. Technology also altered the philosophy driving medical care. Prior to WWII, most medical care was "patient-oriented." After medical technology exploded, health care became "disease oriented."

Disease-oriented medicine emphasizes pathology, differential diagnosis, increasingly complex technology, and aggressive intervention with regard to both diagnosis and treatment. Patient-oriented medicine emphasizes how these may be used to help patients.

It was not just the availability of new technology that fostered this change. In many ways, disease-oriented medicine is far more attractive for health care professionals than is patient-oriented medicine. Diseases are much easier to treat than are patients. Every disease is limited in its scope and the possible ways it can thwart a doctor's attempt to cure it. Patients are not. They typically have more than one disease; they often won't adhere to a treatment program; they live their lives in a variety of different ways; and they are impacted by the wild card of human biology—emotions. These variables can combine in an infinite number of ways to produce an infinite number of problems. *No two patients are exactly the same.*

Diseases also don't talk back the way patients do. It's easier to treat a broken leg than it is to treat a broken leg in the context of its owner. The owner of a broken leg often wants a human/human relationship as a part of the doctor/patient relationship. These relationships are much more complicated, emotionally demanding, and stressful than doctor/disease relationships.

Finally, it's much easier to achieve fame and fortune in a setting that is disease oriented than a setting that is patient oriented. No one ever won a Nobel Prize for taking care of Joe and Shirley and Fred and Gertrude, but year after year, focusing attention on a single disease wins a trip to Stockholm—all expenses paid.

Working to know patients, laboring to understand them, sacrificing time and energy to earn their trust are activities that are not rewarded financially. In fact if done repeatedly they may be penalized financially. But evaluating a disease by putting a scope in a colon or an abdomen, or placing a tube in an artery is rewarded quite nicely, thank you very much.

So it's not difficult to see why a disease-oriented philosophy might be more attractive than a patient-oriented philosophy. Unfortunately, as you may already know if you've ever been a patient, this philosophy has produced health care that is increasingly impersonal and insensitive to patients' needs. It is also ineffective.

An explosion of specialties

Medical technology's explosive growth fostered an axiom that no doctor could know all of medicine. This was gradually twisted into a belief that there may be no role for generalists. The future of pediatricians, general practitioners, and general internists was suddenly in doubt as it became in vogue to think that in the era of modern medicine, urologists should care for all urinary tract infections, orthopedic surgeons were needed for all fractures, and cardiologists were needed for every heart attack.

After a relatively short period of time, problems with philosophy became apparent. For example, consider back pain. This is a symptom that can be caused by kidneys, muscles, nerve roots, blood vessels, female organs, and the intestines. Whom should a patient with back pain go to see? A urologist? A nephrologist? A physiatrist? An orthopedic surgeon? A neurosurgeon? A cardiologist? A general surgeon? Or in the case of a woman, an obstetrician-gynecologist? Obviously, patients needed a doctor who cared for them as people, not walking organ systems.

Realizing that there *was* a demand for their services in the world of modern medicine, general practitioners responded by creating their own specialty, family practice. This was not simply a face-lift for general practice. Family practice demanded a three-year training residency, board certification through an exam, ongoing continuing education, and recertification every seven years (the first specialty to do so). The Academy of Family Practice, the governing body of the new specialty, also worked to define what was appropriate for generalists to do within the scope of their practices. Over time, family practice basically replaced general practice.

The trend in internal medicine, non-surgical treatment of adults, was a bit different. While general internal medicine continued as a source of primary care for adults, it also fostered the development of what were called subspecialties. These began as individual internists who simply had a particular interest in an area like the gastrointestinal system, or the endocrine system and diabetes. Eventually, procedures and knowledge within an area increased to the point where some internists limited their practices to work entirely in that area—they "specialized." What made these specialists different was that they were organized under the general purview of the Board of Internal Medicine—itself a specialty. Consequently, gastroenterology and endocrinology are subspecialties within the domain of internal medicine. To some degree, the same process occurred in pediatrics.

This probably sounds like one of those lectures on "Etruscan Tribes of the pre-Roman Era" that prompts a polite smile and the question, "And your point is?" My point is that because of the emphasis on disease oriented medical care the trend towards specialization had an exaggerated effect of fragmenting medical care and de-emphasizing primary care.

As a new specialty, family practice enjoyed a period of notoriety and attention. Unfortunately, its newness eventually lost its shine. Now all of primary care is once again threatened. Some medical experts will often begin discussions of the future of primary care with the disclaimer "if primary care is to survive in our health care system..." implying that the demise of all of primary care is a possibility. Doctors are leaving family practice and general internal medicine in droves. Younger doctors are not replacing their ranks because primary care has the longest work hours, the lowest pay, and the most intense emotional demands. Primary care is the opposite of disease oriented medicine, and for all the reasons a disease-oriented medical system is more attractive than a patient-oriented system, a young medical graduate has to swallow hard before becoming a general internist or family practitioner.

But without primary care, the entire medical system falls apart. Primary care doctors are the privates and corporals in the war on disease and suffering. No war was ever won without frontline combatants. Every time I hear an expert mention the possibility of primary care not surviving, I conjure up the image of a general hitting the beaches at Normandy, waving his arm and screaming, "Come on men. Follow me!" He looks over his shoulder, realizes he's alone, and promptly gets his backside shot off. So much for the invasion.

If primary care disappears, who will take care of the

respiratory infections, the small lacerations, the bowel problems, the chronic pain, the depression and anxiety, and all the other common day-to-day maladies that comprise the bulk of patients' complaints? It will not be the heart surgeon, the neurologist, the orthopedic surgeon, or the cardiologist. Nor will generals do KP duty or clean latrines. It's not in their job descriptions.

Who will also spend the time with patients sorting through the complicated problems that do not lend themselves to well-reimbursed procedures, the problems that require time-consuming detective work and thought? It is ironic, but over the years I discovered that if I wanted help with a really difficult, puzzling case, I rarely got it from the super-specialists. They were too focused in their one little area of expertise. They also often lacked the ability to listen well enough to obtain a truly accurate history from a patient. In the last ten years of my practice, if I had a really tough case, my consultants were either generalists, or specialists who practiced as though they were generalists.

A doctor, like anyone else who has to deal with human beings, each of them unique, cannot be a scientist; he is either, like the surgeon, a craftsman, or, like the physician and the psychologist, an artist. This means that in order to be a good doctor a man must also have a good character, that is to say, whatever weaknesses and foibles he may have, he must love his fellow human beings in the concrete and desire their good before his own.

-- W.H. Auden

CHAPTER SIXTEEN

HEALERS AND MEDICAL TECHNICIANS

The loss of humanism

The creation of a health care *industry* dehumanized medical care. It changed Mrs. Jones, the mother of three children, who is a die-hard Red Sox fan, and has a Clostridia Dificile infection into "the C. Diff diarrhea in 204 bed 2." It shrunk the doctor/patient relationship into a nonessential amenity. Now, the idea that human beings need extra attention when they are ill or injured is far more an advertising ploy than a tenet of care. "We'd all like a little tender loving care, wouldn't we," is the response I received from one insurance executive when raising this issue.

Humanism is well down health care's priority list. It's below electronic medical records, physician report cards, systems for quality control, and not even on the same page as corporate profits. But what if patients created the priority list? Where would they rank humanism?

Consider these questions. Where would you rank having a doctor who knew you, who honestly cared about you, who had the time to talk to you, and whom you trusted? Where would you rank a system where hospital nurses spent enough time with you that instead of feeling frightened and perhaps even abandoned you felt comfortable? Where would you rank a system that appeared unified in its efforts to take care of you above all else and discounted profit, political gain, and professional grandstanding?

It has been my experience that patients rank these concepts higher than any others. That we do so should not be a surprise. As a species, we have been caring for each other within the context of intimate rituals and relationships for a time that is orders of magnitude greater than the time during which caring has been equated with an impersonal contract. At a very basic level, we expect and need humanism and intimacy as an essential part of our health care.

This is part of the human condition. It can't be ignored. Even the best of the economists, the most erudite of the sociologists, and the most intractable of the politicians can't disprove the qualitative and quantitative equation: one patient equals one person.

Healers and Medical Technicians

I divide doctors into two categories, *healers* and *medical technicians*. There are entire books defining the concept of a "healer" and entire industries based upon its marketability. To mark its difference from the word "technician" is treading thin ice, risking a plunge into the deep waters of philosophy and medical ethics. But my distinction between these two terms is very simple: *Healers care for people. Technicians don't.*

I am not implying that technicians do not offer valuable services. Quite honestly, the "miracle of modern medicine" is far more a product of their efforts than those of healers. Medical technicians are at the cusp of altering the very character of human existence. The profession of healing is not even modern. It's ancient. It has been a part of humanity's culture and social structure for at least 12,000 years. Until recently, healing was closely tied to religion, mysticism, and ritual. As we noted, in primitive cultures, it still is.

Modern medicine has distanced itself from these primitive roots. In doing so, however, it has also distanced itself from the simple truth that as a discipline, medicine's sole function is to care for *people*—individual human beings. People do not regard themselves as simply a collection of organs. Something within them—call it consciousness, a soul, or sentience—declares and defines their humanity. To ignore that particular something is to work as a technician and not a healer.

239

Perhaps I can best illustrate the difference between a medical technician and a healer with an example. I have a friend who is an orthopedic surgeon. He has the inclination to be a great healer. This inclination was best demonstrated during a showdown between a group of powerful HMOs and the orthopedic community. For months, he was literally the only local orthopedic specialist who continued to see all patients with all problems from all HMOs. For the patients and those of us caring for them, he was a godsend. For the orthopedic community, he was a pariah.

Eventually, he left his group to join another—primarily because of his actions during that game of reimbursement chicken. His new group favored super-specialization. He invested some time in further training and began limiting his personal practice to knees and shoulders.

One day I saw a patient of mine following an auto accident. She had a fracture dislocation of a shoulder as well as a back injury. She was obviously in pain, but she was also frightened and angry because an overworked and impersonal emergency room staff had left her feeling that they did not appreciate her pain and the magnitude of her injuries.

Fortunately, I had cared for the patient for some time. The trust garnered from that relationship allowed us to accomplish three things: validate her pain, acknowledge how the accident had disrupted her life and frightened her in a way she had never been frightened, and reassure her that we could outline a course of action restoring normalcy, i.e. return her to a state of *health*.

This is not tree-hugging, warm fuzzies. I suppose it fits into the pigeonhole of the "art of medicine," but categorizing it as such demeans its fundamental role in medical treatment. Addressing the emotions, values, beliefs, and hopes that people carry within them may be a much more subtle task than assessing the imaging studies and lab reports carried within their charts, but it is no less important.

By the time my patient left our appointment, her pain was considerably decreased. Treating her as a person and addressing the emotions accompanying her injury helped her gain control of the pain.

I referred her to my friend the orthopedic surgeon, who was nice enough to work her in that afternoon. That evening, I received a phone call from the patient. She was worse than ever. Her shoulder appeared to be stable, but her back was killing her. I asked if the orthopedic surgeon had suggested anything in addition to what we had already initiated—analgesics, rest, and muscle relaxants. There was a long silence.

Finally, she said, "He didn't even say a word about my back."

I expressed my surprise. She went on to relate that when she had asked about her back, my friend had said that he didn't take care of backs. He suggested that she make an appointment with the back specialist on her way out of the office.

241

"So did you?" I asked.

"Yes."

"What did *he* say?"

There was another long pause. Then, in almost a whisper, she said, "I can't see him for ten days."

My friend had become a medical technician. He had treated a shoulder instead of a patient. The patient underscored that fact with her next question. "Isn't there a bone doctor who will treat me like a person?"

To quote Doctor Robert Coghill of Wake Forest University, a pain specialist, "We don't experience pain in a vacuum. Pain is not solely the result of signals coming from an injured body region, but instead emerges from the interaction between these signals and cognitive information unique to every individual." His opinion is not philosophical. It is a conclusion based upon recent studies using sophisticated brain imaging. It's not a platitude of warm fuzzies. It's hard core neuroscience.

My patient had trusted me. I had helped her establish a belief that the devastation of a bad automobile accident was an event with a beginning and an end. When she called me that evening, she had no reason to invest in that belief. She had left the orthopedic surgeon's office once again feeling abandoned. Her cognitive information—her emergency room experience and her experience with the orthopedic surgeon—implied I had deceived her. Her conclusion was that in truth none of the professionals charged with helping her honestly *cared* about her as a person.

Health *care*.

Medical *care*.

Eventually, I took care of the patient without the services of any orthopedic surgeon. She was fortunate that conservative management was all that she needed. Had matters been different, I think I'd have had great difficulty convincing her to consider surgery. As is often the case, her fear hardened into anger, and her anger evolved into resentment.

When I mentioned these events to my orthopedic friend, he was appalled. His feelings went deeper than just a concern about angering a referring physician. "Good Lord," he kept saying, "I never even thought about that." He had known the patient would have to wait for an appointment with the back specialist. It had not even occurred to him to question what she would do in the interim. Care of a broken shoulder is a much easier task than care of a broken shoulder in the context of its owner.

The trend towards super-specialization has made life easier for medical technicians. It has made life much harder for patients.

A medical technician would observe that not seeing the back specialist made no difference in my patient's conservative management. Had she seen him even immediately, nothing would have changed. A healer would observe that *not* seeing him made things worse. The proof of a pudding is in its taste. That's hardly a medical aphorism, but it is the truth.

Primary care

Medical care used to be delivered by thousands of "Ol' Doc Johnsons." The good doctor Johnson was probably a general practitioner—a "GP." Prior to the World War II, most health care in America was delivered by GPs. The concept of primary care was not really relevant at that time, because GP's were primary care, secondary care, obstetrical care, surgical care, and visiting nurse, all rolled into one. Their training usually consisted of college, four years of medical school, one year of a rotating internship, and a lifetime of experience. This amount of training was sufficient at that time because treatment options were fairly limited.

I entered medical school in 1968. At that time the University of Colorado Medical School was in the process of attempting to transform itself into the "Harvard of the West." That, of course, was never an official policy, but the emphasis on specialization and the support of leading-edge treatments such as liver and kidney transplants testified to its validity. In this environment, GPs were viewed as outdated anachronisms. The community doctors who sent patients to the university medical center were referred to as "LMDs," local medical doctors, and the image associated with this acronym was of a doctor who might not be using leeches and bleeding people but was certainly not "up to date."

As GPs retired or worked themselves into early graves seeing patients, doing surgery, delivering babies, making house calls, and working fourteen-hour days, their ranks dwindled. New medical school graduates entered glamorous and much more lucrative fields such as radiology (x-rays), neurology (brains and nerves), internal medicine (adult medicine), cardiology (hearts and blood vessels), urology (kidney and bladder surgery), nephrology (kidneys but nonsurgical), and orthopedic surgery (surgery of bones). But patients slowed the stampede towards specialization. They realized they had a problem. There were many doctors trained to take care of their bones or brains or kidneys, but there were fewer and fewer trained to take care of them. Who were they supposed to see if they were just plain "sick?" Even more confusing, as we discussed earlier, if they had back pain, should they see an orthopedic surgeon about bones, or a neurosurgeon about nerves, or a urologist about a kidney stone? It seemed silly, but suddenly a patient had to almost know what was wrong with them before they could see someone to fix it. Worse, if they guessed incorrectly there was little chance, for example, a kidney doctor was going to help with an orthopedic problem—"not my area."

It may also be that the cultural revolution of the 1960s created a backlash against the depersonalization of medical care. Most people could still remember "Ol' Doc Johnson" making house calls, and there was some outcry against the

245

dehumanization associated with the emphasis on specialization.

It's appropriate to briefly digress and discuss what being board certified actually means. Specialties have academies or boards that represent and govern them. It is the responsibility of these governing bodies to decide what training and testing are required for a doctor to earn designation as being board certified in that specialty. Seeing doctors who are board certified means you are seeing specialists trained in their given specialty areas. Without this "Good Housekeeping Stamp of Approval" you are not assured of a doctor's level of training in a given area. This is true because doctors can advertise themselves as being experts in just about anything, regardless of their training.

Don't jump out of your chair screaming, "You see! Them damn doctors don't do anything to police themselves!" The reason doctors are not prohibited from what might be called false advertising is first a concern with restraint of trade and second with the fact that specialization has increased as rapidly as it has.

For example, as mentioned earlier, the specialty of gastroenterology (diseases of the stomach, intestines, pancreas, and liver) is actually a subspecialty of the American Board of Internal Medicine. Less than twenty years ago, most gastroenterologists were simply internists with an interest in diseases of the gastrointestinal tract. Now they have branched off into their own subspecialty area. But there are a great many talented gastroenterologists who trained before this happened. It was actually these doctors who developed the field in the first place. Any action limiting their freedom to work in the area they developed serves no purpose. It could also be considered a restraint of trade, placing everyone in a court of law when they should be out taking care of patients with stomachaches from trying to eat the dish they made after watching the Cajun Cook on television.

So, in a world of logic and common sense—PCPs take care of patients for the most part, and when they need help they get it from the appropriate specialist. Right?

Unfortunately, wrong again. This is not a world of logic or common sense. It is a world of self-interest, egomaniacs, greed, and stupidity. Medicine, fueled by a market driven industry, continued to ignore what should be its most basic function, treating people. It continued to evolve into an increasingly fragmented industry of people treating diseases, and everyone trying to make a whole bunch of money in the process.

Primary care should be the heart and soul of medicine. It should be the most prestigious and the most powerful specialty area, populated by the brightest, best trained and most dedicated doctors, and paid on a level commensurate with other specialties. Is this the case?

Sorry. Wrong for another time. A disease oriented, market-driven health care industry has devalued primary care as it has devalued the idea that patients are people. Procedures, technology, a massive growth of pharmaceutical options, and the disintegration of medicine into an ever-increasing number of
247

specialties have overwhelmed medicine's humanism. The quality of a patient's life is often ignored in a pursuit of quantity of years lived. Common sense and clinical experience are virtues only when they meet the ever-changing criteria of "evidence based medicine." At one time, the professional standards and philosophy of doctors and nurses vouchsafed a patient's dignity. Now, hospitals have created the position of "Patient Representatives" to handle the complaints created by impersonal and uncaring treatment.

There is an old saying—"the operation was a success but the patient died." This punch line of a bad joke has become an apt description of American health care. The patient is no longer of primary importance. What is? The system, or the cost, or the statistic, or the treatment protocol, or the kind of medical record, or the billing code producing the highest reimbursement, or the best way of polarizing opinion to gain maximum media exposure, or the best way of hiding profit. None of these priorities favor primary care, because by definition, primary care has one priority—you—the patient. I think it fair to say that as primary care has diminished in stature so has the perspective that the goal of medicine is to serve patients.

This does not mean that all family doctors, internists, or pediatricians are beatific practitioners of the healing arts or even that they are all *healers*. There are good primary care docs and there are some real jerks. But by the very nature of their chosen area of practice, they should be taking care of you, all of you.

They *should* be healers. Ideally, they should know you, understand at least a little about your mechanism of living life, know when to ask for help with regard to specific medical concerns, and, as much as possible, care about you. If they can't do these things, then they are not doing their job. *Primary care should be a specialty of healers.*

As measured by these particular criteria, very few PCPs are now doing their jobs. Why? Because they are not allowed the time to do so; they are not reimbursed to do so; their role is looked upon as being less important than disease oriented specialists; and their job, as I have defined it, is perhaps the most difficult and demanding in medicine. But for a patient, *continuity of care with a PCP is absolutely necessary for good health care.*

For an individual patient, this is the most important thing I will say. Having your own primary care doctor is important because they are the only doctors *trained* to take care of all of you, to deal with you as a person. What is even more important, they are the only ones who have chosen to do so.

To put it a bit differently, in most cases, doctors specializing in areas other than primary care have chosen *not* to care for of all of you. They have chosen to be *medical technicians.* I'm not suggesting that you grab torches and pitchforks and charge the offices of your local cardiovascular surgeon or nephrologist or neurologist. I am not saying that specialists outside primary care are cold, calloused monsters.

There have been a number of studies evaluating why

physicians choose to enter various specialty areas. Sometimes, a mentor specializing in a particular area motivates a young medical student. Sometimes a physician follows a mother or father into a specialty. But more often, physicians drift towards those specialties that match their personalities. Thus, when I was in medical school, my classmates and I were not surprised when the six-foot-five inch heavily muscled former football player, who received as a graduation gift a circular saw from Black and Decker instead of a blackbag from Eli Lilly, went into orthopedic surgery.

Consider, for example, cardiovascular surgeons. Tom Wolfe wrote a book about the original astronauts entitled *The Right Stuff*. At one point, he discussed what it takes to be a test pilot, to have the ability to climb into an airplane for a test flight the day after a fellow test pilot has "augured in" while flying the same kind of airplane. Wolfe said one of the primary personality traits of these pilots was incredible self-confidence and enormous egos—"He crashed because he made a mistake. I don't make mistakes so I'll be just fine." (My words.) He went on to say that the only professionals he knew that had egos bigger than test pilots were cardiovascular surgeons.

I trained with a fellow who was well equipped to be a cardiovascular surgeon. All he lacked was the XXXXL-sized ego. He was destroyed during his training. It doesn't require a Noble Prize in psychology to figure out that in the middle of a heart case, you can't have the surgeon doubting his or her skills or intra-operative decisions. "Ah gee, you know I'm just not sure. You think I put that last stitch in right? I dunno. Maybe I should take it out. Where's that blood coming from? Is it from that last stitch? What about the others? What... er... wait... um... Wow, that's a lot of blood. I mean that's really a LOT of blood. Er...um."

A doctor who has a personality suited to be a cardiovascular surgeon would be terrible at primary care. Taking care of people is a constant effort to deal with problems that are rarely "fixed." Primary care deals with problems that are chronic and compounded by the unpredictable complexities of life itself. Primary care doctors are rarely in control of anything. They must constantly react to Mother Nature's little surprises. A great cardiovascular surgeon excels in precision, decisiveness, and control. A great PCP excels in creativity, patience, and global thinking. You want a test pilot holding your beating heart. You want Doc Johnson, a healer, listening to you try to explain how your new back pain is different from your old back pain and why that has you worried.

A subtle reality ignored by the experts, and many physicians, is how crucial Doc Johnson's ability to listen really is. It is typically called "handholding." This implies that it's nice, but nowhere near as important as what a heart surgeon does. The heart surgeon's work is enveloped in drama, technology, and awe-inspiring images. Doc Johnson just sits there, listening. No self-respecting television executive would ever consider making

251

a television series about that. It would be like watching paint dry. Most of the time it's less dramatic than paint drying—but remember our earlier patient. Dan's problem seemed not terribly acute or life threatening, but it was.

To illustrate my point, let's consider an example from my own experience. Some years back, a patient of mine we will call Harry presented with a complaint of back pain, which he had been suffering from for many years. Chronic back pain can be a terrible problem with recent studies even finding a loss of brain mass in those who suffer from it. (The supposition is that the stress of chronic pain damages the nervous system to such a degree that brain cells literally die.) In Harry's case, the pain was certainly chronic and debilitating, but his wife viewed it as an excuse to avoid taking out the garbage, as well as other household responsibilities. Harry's office visits were often primarily an effort to validate his pain.

When I finally figured this out, I developed an approach that was moderately effective. I would sit and listen to his detailed description of his suffering. I acknowledged his pain, reassured him that I did not think it would get worse, and gave him the latitude to decide what activities he felt he could do. New treatments were limited to therapies that would not make matters worse. Early on, we had done extensive investigations of the pain. He had received aggressive physical therapy, engaged in a monitored exercise program, and even tried acupuncture. Nothing helped.

Taking care of patients who do not get better in spite of comprehensive efforts is a very difficult job. I had no doubt Harry was in pain, but I must be honest. On the occasions when his wife accompanied him to an office visit, I felt as much sympathy for her as I did for Harry.

For those of you with an interest in psychology, the answer is "yes." Yes, the back pain was part of a codependent relationship between Harry and his wife. Yes, I was being sucked into the dynamics of the codependence. Yes, I was aware of everyone's involvement in this process. Yes, I tried to alter its dynamics by referring Harry and his wife for family counseling. And yes, after forty years of marriage, Harry and his wife chose back pain rather than change.

It was against this historical backdrop that I saw Harry late one afternoon. He was my next to last patient. Typically, I saw him about every eight weeks, and it had been only four since his last appointment. The medical assistant's note had referred to "severe back pain," but when I entered the exam room, Harry was sitting in a chair, legs crossed, looking rather bored, but not in any pain.

We exchanged greetings, me apologizing for running late and Harry saying he understood and was happy to have been worked in to see me. He then began a discourse on his back pain. I could have interrupted him and asked that he focus on what was new, but I knew that would be futile. Harry was much better at telling his history the way he wanted to tell it than I was at

253

getting him to tell it the way I wanted to hear it. So, as he recounted a history I knew quite well, I thumbed through the chart to find how long it had been since he had received any physical therapy. I thought that it might have been long enough that his insurance would pay for it. I harbored faint hopes that this time it might help. My search through the chart stopped rather abruptly when Harry used a new word to describe his pain. "Harry," I said. "Slow down for a minute. Did you say the pain was 'searing?'"

He paused for a moment, and then answered, "Yah, searing. That's what it felt like. Sort of hot, tearing—low. Came on real sudden and then after twenty minutes or so, it was gone. Haven't had it since. Course the other pain, oh God that pain is never gone. That's with me..."

"Harry," I interrupted, "tell me again how this pain is different."

He repeated some things, and I probed a little more. It did seem this pain was not typical of his long-standing complaints. Finally, I said, "I want to do a couple of simple tests. Okay Harry? Just want to make sure we're not seeing anything new."

He uncrossed his legs and sat up. "Sure, Doc. Whatever you say. We're not looking at something bad are we?"

"No, I don't think so. But never hurts to look, and these tests are simple. We can do them here."

I paged my medical assistant and asked her to do an hematocrit—red blood count—and a urinalysis. As she passed me on the way to Harry's room, she gave me an eye roll that was less than subtle. She was already irritated that we had worked a chronic problem into the end of the day's schedule. Now, my ordering labs meant she was going to be even later picking up her son from daycare. That not only cost her money, it also ran the risk of antagonizing the daycare provider. Every working mom or dad knows that is akin to angering the gods.

I returned a couple of phone calls while I waited for the lab. My medical assistant brought it back to me, and rolled her eyes again, this time in surprise. "Wow," she said, "Harry's 'crit is down, 29."

"How about his UA?" I asked.

"Normal," she said. "Except for two plus blood."

I'm not sure exactly what sound I made. It could have been "gwak" or "uuuuh" or "shiiiii..." Whatever the sound, it conveyed the appropriate message.

"That's bad, isn't it?" My medical assistant was grimacing. She's a compassionate person, so I know the grimace was not only because she could see the possibility of an even later end to the day but also because Harry's problem had taken on the specter of something worse than chronic complaining.

"Harry might have a dissecting aneurysm." I felt like adding *"and there is the distinct possibility he could blow the thing out and bleed to death any minute."* Instead, I drew in a breath and considered possibilities.

I decided the best option was to send him to the

255

emergency room. What he needed was an ultrasound of his abdomen, and the chances of me getting that done as an emergency outpatient procedure were nil. Further, if he did indeed have an aneurysm, he needed to see a vascular surgeon, and maybe ASAP. I called the emergency room, talked to one of the ER docs and filled him in on Harry's history. Then, I left a message for a vascular surgeon I knew was not only good but would also see Harry that evening.

I returned to Harry's room. He was waiting with a very apprehensive look draped across his long, lined face. "Harry," I said, trying to strike a happy medium between reassurance and doom, "it looks like we'll need to follow up a bit further on this new pain of yours. You're going to hate me, but I've got to send you to the emergency room."

"No way! I can't go there, Doc. Not only is that place a zoo, the co-pay is a hundred bucks. My wife would kill me. Doc, I don't have the pain anymore. It's gone."

"Harry, your wife would not kill you. She's much kinder than that, and I've watched you in close combat. You handle yourself okay. She won't kill you, but I want to make sure whatever caused your pain doesn't kill you either."

He dismissed my concern with a wave of his hand.

I raised my voice a few decibels. "Here's how it is, Harry. Either you agree to go across the street to the emergency room or I pick you up and bodily carry you. I'll do that, but if I hurt my back, and there's a good chance I will because I'm old and out of shape, I'll sue you for everything you're worth—all seven dollars and sixteen cents."

Harry turned his head and looked at me through one eye almost squinted shut. "You're serious, aren't you?"

I nodded.

"What do you think I could have?"

I mentally breathed a sigh of relief because I could see I finally had Harry's attention. "I'm worried about an abdominal aneurysm, a weakness in the big blood vessel that comes out of your heart. People can be born with a weak spot that balloons up over time. Finally it becomes so weak that it ruptures."

"Ruptures?"

"Right. If it breaks in the part that's open to the inside of the abdomen, it's like being shot with a bullet. You bleed to death inside. If it breaks backwards, it bleeds for a while, but there's not much place for the blood to go because of the muscles and the kidneys and everything that's back there. The bleeding stops. That's why we did your red blood count. It shows that you've probably bled somewhere. I think you had a bleed backwards called a partial dissection. You been feeling weak?"

"Doc, I'm always weak."

My medical assistant wheeled him across the street to the emergency room. He probably did not need to be in a wheel chair, but it served to make sure he got there without going home for clothes or a toothbrush—a decision that had burned me in the past. My next task was to clear the entire process with his

257

insurance, something I theoretically should have done before sending him directly to the ER.

If you're not a primary care doctor, "coordination of care" is probably a pretty abstract idea. It is one of primary care's more difficult and thankless jobs. For me, it involved making sure that patients received appropriate care by referring them to specialists I personally knew and trusted. It also involved overseeing patients' treatments. Specialists were often unaware of the entire scope of patients' problems, and this raised the possibility of the right hand being oblivious of prescriptions written by the left. Finally, coordination of care sometimes included having to help patients maintain their identities as human beings when they felt overwhelmed by multiple doctors and endless procedures. None of these tasks were paid for on any consistent basis.

In the last dozen years, coordination of care has become infinitely worse. Insurance companies, or in the case of Medicare patients the federal government, have established an ever-increasing body of rules, statutes, and stipulations. Coordination of care has become a version of "mother may I?" played with a phone clerk. Before considering any significant medical treatments, doctors must play the game of begging to appropriately care for their patients.

Often I can also no longer assure my patients of the competence of a specialist because whom they see is determined not by my trust in a specialist's skills and judgment, but by whether or not he or she has signed a discounted contract with an insurance company. Consequently, I must refer to specialists I don't know.

Just knowing who is and who is not on a specific insurance company's panel can be a mind-numbing proposition. At one time, our practice had a full-time employee just to handle referrals. Over the years, she became almost encyclopedic in her knowledge of which doctor took which insurance and who to call if there was a problem. When we had to eliminate her position because of decreasing reimbursements, we were all devastated.

So, on the evening I saw Harry, having made a tentative diagnosis, having convinced the patient to go to the emergency room, having apprised both the emergency room doctor and vascular surgeon of the patient's status, I faced one more task—calling the insurance company.

Harry did end up having an abdominal aneurysm. It was still leaking, and he was operated on that night. On hospital rounds the next morning, I happened to enter Harry's room at the same time as the surgeon. Harry's family had spent the night, and they were eager to find out how he was doing. It was apparent the surgeon held their rapt attention. They hung on his every word, all six of them. "Doing okay. Blood count's stable. Seeya."

As the family tried to coax a few more words from the surgeon—had he gone beyond a dozen it would have qualified as a soliloquy—I slipped out of the room. No one seemed to notice. That did not bother me, because I was already late for the office. I also knew that soon enough I would be asked to utter profound statements about Harry's future as the acute emergency evolved into the ongoing routine of day-to-day life.

In this particular battle in the war on disease and suffering, there was no doubt the heart surgeon was the test pilot, the jet jockey. He walked into Harry's room wearing his surgical greens and a heavily starched long white coat with a surgical mask draped around his neck. I walked in wearing a shirt and a Mickey Mouse necktie given to me by a patient coming in for a six-year checkup later that day. I was definitely a corporal, an infantry grunt. But I knew I had done my job well. It may have been the skilled hands of the surgeon that resected Harry's bulging aorta, but it was my ability to listen and think, the frontline work of medicine, that had granted him his chance at a successful surgical intervention.

A difficult but not impossible job

It is ironic that patients will often adore their own doctor without ever really appreciating how difficult a job primary care is. I once had an honest, insightful cardiologist say, "Man, you got the toughest job I know. I really only have to know eight diseases. But you have to know them all. And you never know what's on the other side of the door." He's right—sort of.

If primary care doctors needed authoritative knowledge of every disease and treatment to effectively care for patients, nobody could do the job. The assumption that such knowledge is necessary to practice primary care is sometimes used as an argument against its viability. (It's not a valid specialty because no one can do it in a competent fashion.) But primary care docs do not have to be authorities in every disease. They need to be an authority in you.

That's the sort of statement that makes a great sound bite and fits well on a brochure. But what does it mean?

It means that the most important information relating to any patient is how they are doing. "Howya' doin'? What's new? Problems with your meds?" are very simple questions, but if a doctor knows the true answers to those questions, he or she has a fairly good handle on a patient's status. The operative word in that last statement is "true," because as we've discussed, there are barriers to patients revealing the truth, and there are barriers to doctors hearing it. It often requires a skilled specialist in *people* to obtain medicine's most crucial data—Howya' doin'? What's new? Problems with your meds?—regardless of whether the patient has a cold or a really rare disease requiring the assistance of a specialist in really rare diseases.

At one time, the ability to pry pertinent information from a patient and put that information together into the pattern of an accurate diagnosis were considered two of medicine's greatest skills. Modern medicine's early heroes, like Sir William Osler, were all very proficient in taking full, accurate patient histories. This skill combined with a meticulous physical examination comprised the major part of a patient's initial evaluation. This was still true when I was a medical student, forty years ago.

Medical Metamorphosis

While all medical students are still drilled in the details of doing physicals and taking histories, there is not nearly as much emphasis placed on their value. Now, laboratory tests and imaging procedures are considered of equal if not greater importance. This is somewhat appropriate, because as technology has exposed the inner workings of the human body, it has become apparent that some classic physical "signs" don't really mean much. It is probably true that an axial CT scan can reveal as much if not more than a very detailed physical exam, and it certainly holds more credence in a court of law. However, this reliance on technology in lieu of old-fashioned skills is a trend that not only further dehumanizes patient care, it also carries significant cost and risk. There are many situations that do not need CT scans or MRI's. What they need is the attention of a good PCP. They need the attention of a healer.

I must emphasize that I am not promoting incompetence in favor of empathy. A healer must be professionally skilled and knowledgeable in addition to being able to relate to patients humanistically. Doctors who can relate but are incompetent are what I call "back-slappers." They are often successful because they are gifted in the art of reassurance punctuated with a friendly slap on the back. They are also dangerous.

Once again, let me demonstrate with an example. During my rotation through the intensive care unit when I was an intern, my internist friend and I cared for an exceedingly ill patient. His attending physician was a well-established and much beloved general practitioner. Unfortunately, this doctor had not read a professional journal since finishing his internship forty years earlier.

Two days after the patient had been admitted to intensive care, this doctor showed up to see him. The patient was obviously overjoyed to see his doctor. He even tried to sit up. His doctor talked to him briefly, his hand on the patient's shoulder in a reassuring gesture throughout the conversation.

The doctor then came over to the physician's station and picked up the patient's chart. He looked at it briefly and wrote a note of his own. I started to give him a report on the patient's progress, but he smiled and said he was late for his office. His entire visit took less than five minutes.

I picked up the chart and looked at his note. My note was a page and a half long, describing each of the patient's problems, listing their relevant data, and offering an assessment of their status (what is called a problem-oriented or SOAP note). The length reflected the serious nature of the patient's condition. He had almost died, was on multiple intravenous medications, and still not out of the woods. My internist friend, who at that time was my resident, had written an equally lengthy note. There were also many pages of orders.

The attending physician's note read, "Deaf but alert."

For me, those three words became a symbol of back-slapping. For physicians to be healers, their professional skills must be equal to their humanism. Having a relationship powerful enough to make a gravely ill patient try to sit up is but half of a physician-healer/patient contract. The other half is professional competence. In the context of this particular patient, a progress note that read, "Deaf but alert," hardly implied competence. It implied an indifference to standards of care. It defined the doctor as a back-slapper.

Back slappers are a danger to both their patients and to healers. The danger to patients is obvious. Fanny slappers are medically incompetent. In the vernacular of interns and residents, "They screw up all the time."

They are a danger to healers because back slappers' humanism becomes synonymous with their professional inadequacies. I have listened to experts berate practitioners for their humanism, their attempts to show compassion or develop relationships with patients, and the criticism often ended with the admonition that time spent in these efforts would be better spent, "making sure you know what the hell you're doing." The dangerous implication of this criticism is that humanism cannot go hand in hand with medical proficiency. In this regard, back slappers have maligned the crucial role that healers play in appropriate medical care. It has given fodder to the assumption that in a world of complex treatments and demands for almost instant diagnosis, the job of a PCP, a healer, is not possible.

But this is not so. As I said, the most important database is *your* database, the patient's. The most important knowledge is an honest understanding of a patient's status. Further, a development has made it possible for PCP's to be instant experts in almost any problem—the Internet search engine.

In the last five years of my practice, my favorite consultation was from Doctor Google. If I faced a problem that required knowledge greater than what I had stored between my ears, I went to the Internet and used Google™ or one of the search engines specific to medical information. Within a short period of time, I had the most up-to-date information about that problem.

I exaggerate when I use the term "instant expert." That suggests that *experience* is of no importance—it is. Even if I found a solution to an unusual or complex problem, I more often than not still sought the advice of a specialist. But I did so armed with an up-to-date understanding of treatments and diagnostic techniques that before would have been impossible.

Given their training, the requirements for their ongoing reeducation demanded by their specialty boards, and the new sources of immediate access to unlimited information, modern PCPs have the potential to deliver unparalleled medical care. They have the potential to integrate the miracle of modern medicine and the art of healing.

However, it well may be that this remains only a potential. It well may be that primary care as a specialty ceases to exist within a decade.

A dying breed

Now that I've emphasized the importance of having your own PCP, I need to let you know that if you don't already have one, you'd better start looking. PCPs are leaving the specialty faster than anyone wants to admit, and few medical students are choosing primary care as a specialty. I practiced medicine in Aurora, Colorado, a suburb of Denver. Seven years ago, in a fourteen-month period, Aurora's primary care community decreased by eighteen percent. There were one or two retirements because of age, but most losses were well-established doctors saying as I did—"enough." One popular female internist completely left medicine to work with her husband at a fast-food establishment. Her explanation was quite simple. "I would rather flip burgers than go through what I've gone through for the last two years." When she returned to practice, she had retrained and completely left primary care.

The Denver Post carried a story about a medical student from a rural Colorado town. He had always dreamed of being that small town's Ol' Doc Johnson. In summers while he was in college, he had worked in the office of the man presently filling that need. The community was anticipating a passing of the baton and took pride in what appeared to be a legacy of care.

The story related that the legacy would never come to pass. The medical student had taken a look at his growing student loans and the level of reimbursement for PCPs. He had realized what sort of hours were going to be required to deliver care and the toll those hours would take on his family and himself.

He had applied for training as a dermatologist. The small town would not benefit from the skills of its native son. The young medical student had abandoned his lifelong dream.

Much of the pressure driving doctors from primary care is financial. Twenty-five years ago it was virtually unheard of for a doctor to go "belly up," for a practice to completely fail financially. It's not uncommon in 2006. It's not just bad doctors who fail financially, either. One of Aurora's financial casualties was The Colorado Family Practitioner of the Year.

Overhead for a private doctor has increased well beyond inflation every year for the past six. Reimbursement, on the other hand, has decreased anywhere from five to seven per cent per year. That is a huge net change over the span of five or six years. The only way a primary care doctor can accommodate to such change is to increase volume—to see more patients. This becomes a terrible spiral, because as more doctors quit their practices, more patients look for new doctors. The remaining doctors must then see more patients. The insurance companies cut reimbursement further. Doctors are faced with increased financial pressure while working longer hours, and ... There's no evidence that this spiral will stop any time soon.

In 2003, for the first time, family practice training programs did not fill all their slots for residencies. Young medical graduates just don't want to work increasing hours for decreasing pay while the legal profession looks over their shoulders, insurance companies devalue their services, and SEC officials threaten criminal charges if they band together with other doctors to negotiate with insurance companies (at the same time insurance companies like Anthem take over smaller companies and grow into multibillion-dollar entities).

Gee, I just don't understand why they wouldn't want to go into primary care.

Continuity {of care} implies a sense of affiliation between patients and their practitioners (my doctor or my patient), often expressed in terms of an implicit contract of loyalty by the patient and clinical responsibility by the provider. The affiliation ...fosters improved communication, trust, and a sustained sense of responsibility.

-- "British Medical Journal", 2003; 327, 'Continuity of care: a multidisciplinary review'

CHAPTER SEVENTEEN

CONTINUITY OF CARE and TRUST

Continuity of care

A long-term doctor/patient relationship is one of medicine's greatest therapeutic and diagnostic advantages. The one thing a patient can do to assure good medical care is to establish a relationship with a trusted PCP and then do whatever it takes to protect that relationship. *Continuity of care with a primary care physician is absolutely necessary for good health care.*

More times than I can remember, I have walked into an exam room, looked at a long-time patient, and within three or four minutes initiated his or her transfer to the hospital or the emergency room. I could often tell how sick long-time patients were after just looking at them. I don't say this with any false pride about my clinical skills. I think most experienced clinicians would say the same thing.

In some cases, continuity of care can make a difference between life and death. Years ago, I read an article by a computer programmer who was working on a program of clinical algorithms. These are a series of steps a doctor takes as he or she starts with a patient's general symptom, like "my stomach hurts," and continues to define it and narrow it down to an exact diagnosis with hopefully an exact treatment. An algorithm is like a tree with the general symptom at the base of its trunk and the final diagnosis and treatment in the uppermost branches. To get to the tree's top, one must climb through a series of branching decisions, with one leading to the next. The computer programmer was evaluating the theory that with the use of algorithms, computers could do just as good a job of diagnosis as could doctors.

The author of this article related working with the chief of emergency medicine at a busy urban hospital. One day they strolled through the emergency room waiting area, and mid-sentence the doctor suddenly stopped talking. He excused himself and quickly walked to a patient awaiting treatment. He talked to her for a few seconds, and then with an obvious sense of urgency, escorted her past the admissions desk into the exam room area. He commandeered an exam room and grabbed one of the doctors who immediately began seeing the patient.

When the emergency room chief returned to the

programmer he apologized. He said the woman had been about to go into shock. The programmer was surprised. He had not thought the woman looked particularly ill. The doctor explained that he had seen this patient before, and when he noticed her sitting in the waiting room it was quite apparent to him she was very ill.

The computer programmer's article concluded that this particular episode had proved to him that there were certain skills no computer would *ever* have. He postulated that the software inside an experienced doctor's head has access to data like facial expressions, body language, a patient's demeanor, and a thousand other cues the programmer probably could not even imagine. He ended his article by suggesting that this sort of database could be found nowhere else *other* than inside an experienced doctor's head. With regard to our metaphor of a tree, this gives the gift of flight to a doctor treating a truly sick patient. Rather than laboriously climbing from a set of symptoms, branch by branch, to a diagnosis, and then to a treatment, a doctor who knows a patient can soar from the base of the tree to its uppermost branches. This can be of critical advantage if time is a factor in the patient's care, like during the medical emergency of impending shock.

I remember being gratified by the programmer's conclusions. At that time, which was almost thirty years ago, the proposition that computer programs could replace doctors was a popular point of discussion. It was perhaps the first in a series of challenges to the idea that patients actually need their own physician. I had not been in practice for very long and while I was interested in cutting-edge medicine, I had just taken out a sizable loan to set up my practice. My collateral for the loan was the seat of my pants. I had also just been blessed with the birth of my first child. The idea that a computer could replace me was far fetched, but my new responsibilities rendered me vulnerable to far-fetched threats. I recall a computer geek commenting that an "Apple" (computer) a day might truly keep the doctor away. He was shocked when I critiqued his sense of humor as being "moronic, sophomoric, bereft of social value, and generally swinish." He thought he'd been pretty clever.

In the computer programmer's article, the emergency room chief's general clinical experience was augmented by the good fortune that he had seen the patient before. The woman must have had a guardian angel with her that day. A doctor who knew at least a little about her serendipitously noticed her while she waited, almost in shock, to see someone else. The guardian angel managed to create a bizarre kind of *continuity of care* and probably saved her life. Unfortunately, the medical industry is far too reliant on guardian angels and far too ignorant of appropriate clinical concepts—like patients having their own doctors and enjoying continuity of care.

Another office visit

Let's further explore the importance of having your own

doctor and continuity of care by looking at a situation less dramatic than that of the woman in the waiting room. I want to put you in the shoes of another patient. These are size six running shoes owned by Linda, a thirty-six-year-old mother of three who also runs her own restaurant, a responsibility she shares with her husband. You are Linda. You are visiting your doctor because of back pain. Your doctor has cared for you since you were fourteen years old. He cares for your children and your mother as well. He is always busy, and you're glad you get to see him today instead of one of his partners or one of the physician assistants.

This is a circumstance much less acute than the one the computer programmer wrote about. It would seem that continuity of care might not be as important in this less-acute setting. Let's see.

The door opens and he walks into the exam room. He's running late as usual. You know his tardiness is because he refuses to race through appointments, and it is a predictable part of seeing him.

"Linda, my favorite chef, female athlete, mother of three, and long-suffering wife of the grouchiest man in the world. How you doing? Not great or you wouldn't be here. Sorry I'm late."

You smile. "It's okay, Doc. You're a busy man."

He sighs. "That's no excuse for me taking your time. You're busy too. I just can't find that fifth gear anymore. I'm like an aging point guard who tries to run the fast break and can't."

You continue to smile, realizing he uses this example because you coach a women's basketball team. It's one of his ways of relating, but you still enjoy his acknowledgement of this part of your life. Your smile fades a bit as you think of your husband. "Speaking of the grouch how is he, Doc? That chest pain scared the poop out of me."

Doc sits down on the rolling stool. He brings it imperceptibly closer. "Laura, Dave will be okay if he takes that stuff I gave him to cut down his esophageal reflux, and if he cuts way back on the booze."

You feel a sudden twinge somewhere between your stomach and your heart. "I know, Doc. I know. It's just that..."

The Doc leans over and pats your knee. "He and I talked about it. If he hasn't talked to you about it in a week, call me. But no more about Oscar the Grouch. It's your appointment. Tell me about the back pain."

You describe the pain, a dull ache that grows on a daily basis. It began on one side but now seems to hurt from your butt all the way up to your neck. As you describe the pain, Doc looks at you intently. From time to time he grimaces as you describe how the pain's intensity has increased.

When you finish, he nods. "All right madam,

stand up for me please. Which side did you say it started on?"

"The right." He reaches around your back and with two fingers presses on a spot. The pain almost buckles you. "Damn Doc! That hurt!"

"Sorry. Sorry. Man, you are really tender. That's your sacroiliac joint. It's where your spine articulates, comes together, with your pelvis. My dear, it is red hot, really inflamed."

"I guess!" You rub the hot spot. "What did I do?"

For a moment Doc looks at you. Then he says "I'm guessing the basketball season just started. I'm also guessing you're cooking about ten hours a day at your restaurant. I'm also guessing you're running about five miles a day so you don't return to the cherubic form you had when you were in college. Am I right?"

"Pretty much." You rub your back. "You are really good. You found that thing with NO trouble."

"That's because of my amazing clinical skills and the fact that you had the same thing a few years ago."

You search your memory. "Oh, right. You remembered that? I didn't even remember it."

"Linda, it's amazing how patients put the memory of painful events somewhere in their brain just next to the memory of their worst high school date. They don't *want* to remember it. I remember you having this because last time it was even worse. You came in here about eighteen inches tall, walking sort of like a crab."

You laugh. "Now that you put it that way, I'm beginning to remember all too well. How do I get rid of this thing?"

He shakes his head. "I'm sorry, but you have to give up coaching, stop running and control your weight with just a diet. You also need to hire a cook to work your shift for a few weeks." He pauses. Your shoulders drop. "And.... I want you to take up Zen meditation, gradually increasing your meditative sessions to three or four hours a day." He closes his eyes and assumes an exaggerated meditative pose. "OMMMMMMMMM, oh ya, OMMMMMMMM."

You both laugh. "What?" he says with an exaggerated shrug. "You don't think that will help? Well, if you're not willing to help yourself..."

There is a moment of silence. Then he says, "You can't afford a cook, and you sure as hell aren't going to stop running or coaching, right? They're the only things keeping you from going bonkers."

You nod.

"Well then, we gotta find some way to cool off that joint other than Zen."

He explains a little bit about what inflammation of the sacroiliac joint does to muscles, how running and its jarring impact make things worse. He digs through a drawer and finds a sheet of stretches and reviews how they should be done. He discusses the dose of an over-the-counter anti-inflammatory medication and how icing the joint after running might help and how even trying swimming can help until your back improves. He explores ways you can take breaks while you work and how standing in one position can also make things worse.

As you are about to leave, he drops his voice. You know he's about to say something important. "Linda, the other thing that makes it worse is general muscle tension. Look how you're carrying yourself. Absolutely rigid. You look like a guard at Buckingham Palace." He places two fingers on your right shoulder. "Drop it down. Let it relax."

You let your shoulders sag a bit and realize he's right.

"Kiddo, you're a type A plus, and you're never going to change your stripes to spots, but everything you're trying to do is taking its toll. And Dave -- he's not helping."

Tears begin to well up in your eyes.

"We can joke all we want about his being a grouch, and I know he's a good man. But I suspect he's drinking more than either of us knows. And I suspect you suspect that too."

There is another moment of silence.

"I also suspect that if I don't stop my suspicions, you're going to start crying which will embarrass you to no end. Look, if Dave hasn't mentioned the drinking in a week, well, then we're going to have to do something else."

You swallow hard. "Thanks Doc."

"Hey," he says in a big voice, "Like I said, you're my favorite chef-slash-coach-slash-mother-slash-runner-slash-patient."

As you leave his office, you begin to feel better. You begin to feel hopeful.

Now, let's move forward six months. You're still in Linda's shoes. You're about to go to another doctor's appointment, but it's with a different doctor. Because of cost, you've had to change health insurances, and Doc is not on their "list." It was your husband Dave's decision to make the change. You fought the decision, but the economic realities of operating a small business eventually prevailed. Your back pain has returned. Dave is also drinking again. Doc had finally convinced him to go to Alcoholics Anonymous, and for a time Dave's drinking had stopped. Now it's started again, heavier than ever. He refuses to see the new doctor.

The new doctor is female, recommended by a neighbor. She is part of a ten-doctor group that practices out of a clinic owned by a local for-profit hospital. As a patient, you've not seen her, but you have met her once when you took one of your children to her for a well-child check. She seemed pleasant enough and related wonderfully to your child. Waiting in the exam room, you're still apprehensive. This is the first time you've seen anyone other than Doc since you were fourteen years old.

You are Linda.

The door opens, and the doctor enters the exam room. She wears a white coat, something Doc never did. "Hello," she says. "I'm Doctor Paquette. You must be Linda."

You smile a very large smile. "Yes. I'm Linda. I met you when you saw my son at his annual checkup."

Doctor Paquette nods but obviously does not recall the event. "I'm sorry if I don't remember. I just started here four months ago, and all the patients are new. What can I do for you?"

"It's...um...my back, Doctor. I have back pain."

Doctor Paquette reviews the patient history form you filled out prior to the office visit. It was mailed to you at your home when you made the appointment as a first-time patient. For a moment the doctor says nothing, pouring over the five pages of your form. Finally she looks up. "I'm sorry. I was just trying to learn a little about your history. It says here that you've had back pain before. Is this pain anything like that?"

You draw a deep breath. "Yes, I guess so. But worse."

Doctor Paquette checks your history form by asking some questions and then does a thorough exam. She has you move through a series of ranges of motion, inspects your back for scoliosis, and tests the strength of your legs. You think about pointing to the spot Doc found with his two magic fingers, but for many reasons you do not. When the exam is over, Doctor Paquette has still not found the sore spot.

You reach around and gingerly place your finger on the spot. "It's most sore here, Doctor Paquette. My doctor, I mean my prior doctor, said I had a hot sacroiliac joint."

Doctor Paquette pokes at the area. You wince and withdraw. "I see," she says. "Tell me about it. Tell me what he did."

Slowly, you relate your story. The doctor does not interrupt until you describe your running, coaching, and ten-hour days. Then she raises her eyebrows. "Wait a minute Linda. You're still doing those things—with the pain?"

You squirm a bit. "Well, yes. I have to."

Doctor Paquette shakes her head. "No, Linda. You don't have to. We all have choices. Who are you trying to impress?"

You feel the muscles in your neck tighten. Your back pain becomes worse. "Doctor, this is just how I live my life. I work. I take care of my children. I run and coach. That's my life. The back pain, well it..."

Doctor Paquette interrupts you. "Linda, I'm sorry. Our time is up. I don't want to fall behind or the administrator will fry me."

She smiles. You try to smile back.

"You need to set up a complete physical," she says as she writes a prescription. "It will take a little time to set that up, but in the interim, here's what you do. No more running. I know coaching will be hard to stop—you don't want to fail those kids—but see what you can do. We will need to talk about work later on. We will also need to talk about you trying to do too much. Here's a prescription for an anti-inflammatory and an order for an x-ray of your back. If there's anything on that, I'll let you know."

The doctor looks up expectantly, cocking her head to one side as though encouraging any other questions. She does this as she places her hand on the doorknob. You wave good-bye, and she leaves quickly. Your back now aches terribly. For a few moments you sit staring at the x-ray order and prescription. Then you stand and square your shoulders.

You're almost home when you realize you said

nothing about Dave's drinking. You shake your head as you realize you could never have said anything about Dave's drinking.

If I were to write a postscript for Linda's two office visits, it might be fairly painful. Dave, Linda's husband, might continue drinking and the family could be disrupted. Linda might be unable to regain her equilibrium and sink into a depression. There are many bad outcomes, but I find them too painful to create, or relate. It is also possible that Linda finds support from other family members or that she works her way through the awkward period of becoming acquainted with her new doctor and finds the doctor to be a resource, a healing force within her life.

The power of the relationship with Doc is obvious. He knew Linda and her family, and this knowledge allowed him to treat her within the context of not only her pain but also within the context of who she is—both to herself and to the other members of her family. He knew her well enough to have some insight, learned over time, about the mechanism of her life. This insight allowed him the ability to treat her within that context of her life and identity rather than by an abstract set of parameters. It is impossible to overestimate the power of a doctor respecting a patient's sense of priorities, lifestyle, and ethics.

Because Doc was also her husband's doctor, he was in a position to help Dave with his battle against his alcoholism. In so doing, Doc helped strengthen the entire family unit. This was a doctor/patient relationship with a maximum potential to heal a maximum number of people.

Linda's new doctor, Doctor Paquette, appeared to be well intentioned. She also was attentive enough to immediately understand the role her patient's overachieving personality played in her physical complaints. However, Doctor Paquette was caught in the cogs and gears of "doc in the box" medicine. This is an environment not conducive to continuity (and perhaps quality) of care. In this environment, schedules and their maintenance, productivity and its measure reign supreme; physicians work shifts, and when their shift is over their responsibility is over as well; longevity and stability of practice are rare; most encounters are first-time or second-time encounters; and, an understanding of who a patient is—at best—comes from a medical chart, a collection of clinical descriptions that can never impart a true understanding of a human being's self. Thus, the chances that Linda will establish a healing relationship with Dr. Paquette are probably slim.

Trying to mesh with the cogs and gears of her work, Doctor Paquette fell prey to the temptation to confront her patient with "instant insight." Recognizing Linda's overachieving and type A personality is an acknowledgement of Linda's humanity, but telling her to completely change the way she lives her life is not only unrealistic, it also discounts Linda's belief system. Linda perceives this as a direct attack on her personally, as would most patients. This often drives them deep within themselves, a place inaccessible to any doctor's efforts.

283

As I have said, often times the comfort that comes from a long-term relationship with a doctor is discounted as another nicety, a warm fuzzy. Thirty years of experience proved to me that it's no warm fuzzy. It's a linchpin of appropriate care.

Trust

Health care has suffered greatly because the experts have ignored something that in its most general sense pervades every action we take—trust. When you sit down in a chair, you trust it will not crumble beneath your weight, sending you to the floor and endangering your keester. There was a time when a person either relied on the soundness of a chair because they had made it or knew the person who had. If you knew that a master artisan had made your chair, you could plant your backside with a relative degree of confidence. On the other hand, if your brother-in-law had made the chair just before he was booted out of his furniture maker's apprenticeship because of gross incompetence, you might choose to stand rather than sit. Trust in the chair was a personal proposition.

Now, that is almost never so. If you are seated, have you any idea who made the furniture upon which you sit? Maybe—if you're in your own home. You may have purchased that chair because its manufacturer has a reputation for making furniture that does not commonly collapse. But reputation is not a personal experience. It is the product of other's opinions and advertising. Maybe you purchased the chair because you owned one just like it and it withstood a stress test administered by Stanley, your 325-pound next-door neighbor. You're still unlikely to know the person or persons who glued it together. The trust implicit in sitting down in a chair is a much more depersonalized experience than it was three centuries ago. So what? Who expects to have a neighbor make their furniture? It's much better to have Stanley as a neighbor because he works for Krispy Kreme and brings home leftover doughnuts.

Let's increase the complexity of the act of trust. You step on an airplane to go on vacation. Now the stakes are higher. Falling from 30,000 feet produces more than bruises on one's backside. Trust in an airline is also immensely more complicated. You are relying on the skill and integrity of the pilot, the thoroughness of the airline mechanics, the intelligence of the aeronautical engineer who designed the airplane, the soundness of the company that made the airplane—the list is not endless but close.

This is an ultimately depersonalized experience. It has to be. It's too complicated to be otherwise. Society has therefore labored to endorse the *reputation* of this experience with *systems* designed to evaluate those areas of trust. Pilots must have licenses. Aeronautical engineers must have degrees. The Federal Aviation Administration is a government entity whose sole purpose is to guarantee a trust that traveling in an airplane is a safe experience.

But this is still a depersonalized experience, and some

people are incapable of trusting it. Fear of flying can be caused by many factors, but an absence of trust in abstract *systems* is one of the most powerful. That is why one of the approaches to soothe this fear, which in today's society can be debilitating, is to humanize it. White-knuckle flyers are introduced to the pilot of their plane. Flight attendants will spend a little extra time coming to know an apprehensive passenger's name and a little about them. Anxious passengers are always seated either next to a window or on an aisle seat, whichever makes them more comfortable. Right?

Wrong. Surprised? Instead, how about cramming a passenger into a seat wide enough for someone who weighs ninety pounds, whipping a pack of peanuts containing three peanuts at the passenger's head, laughing at a request for a pillow or blanket, and charging out the wazoo for a can of pop. That is the present approach to airline passengers, and attention to one with a fear of flying is usually limited to a shrug and an offhanded, "Sweetie, maybe you should have taken a train." The bottom line for the airlines has become an elusive goal, and for most airline companies, unless you are willing to pay a lot more for a first-class ticket, depersonalization is the order of the day.

Yet fear or no fear, trust or no trust, most of us continue to fly. It's part of modern life, and often we have no choice. Because we're forced to trust the system, we do. Psychologically that's a lot easier than the alternative and certainly less upsetting than taxiing down a runway in an airplane with half the passengers screaming, "I'm too young to die!" We suck it up. We may be apprehensive, but we smile at each other while we grind molars into plaque dust.

In an age of much-publicized terrorism, the powers that be have complicated airline travel by adding many layers of security that are also to be trusted while emphasizing that we should *not* trust our fellow travelers. For me, this has often proved to be a counterintuitive process. I've found it much easier to feel comfortable with a total stranger as we stood in lines hundreds of yards long and discussed the typical issues that strangers standing in lines discuss than to trust the critical acumen of a security screener who sat staring at the monitor of an imaging device with a look so vacuous that he could easily have been dead for days or weeks. I still got on the airplane.

As society has become more complex, we have accommodated to the abstraction of trust. In the case of commercial airlines, statistics support a trust in its systems—or more honestly, statistics support a trust in the *people* who implement its systems. It's considerably safer to fly from New York to Los Angeles than to drive to the local supermarket.

What about medicine? Abstracting trust won't work with health care as readily as it does with the airlines industry because an airline pilot does not ask you to take off your clothes, climb into a paper gown, and sit waiting in a cold room until he or she enters it to ask you questions about the most intimate aspects of your life. Doctors do. They also then proceed to infringe upon

287

even more intimate parts of your anatomy. The most intimate interaction you have with airline pilots is when they announce their names over an intercom and describe the weather in the city of your destination. (They may also describe some point of interest that is visible only from the side of the airplane on which you are not seated.)

The intimacy of the medical experience demands a level of personal trust. When you have mustered up the courage to tell a doctor about some secret concern you have about some embarrassing physiologic function or body part, you don't give a damn if the system that the doctor works for just received a glowing review from the National Committee for Quality Assurance (NCQA). You want a doctor who knows you, cares about you, and one you trust will not burst into laughter when you finally reveal your secret.

The experts have suggested that personal trust between doctor and patient is a failed concept and that it should be replaced by that of an "educated consumer." This concept proposes that patients research physicians, procedures, and medications the same way they research the purchase of a television set or automobile. Further, they emphasize that cost be factored into this decision-making process.

Obviously, for this concept to work, there must be some sort of "Consumer's Report" with ratings of physicians and procedures, but there is some dispute as to what criteria should be used in these ratings. One suggestion is that physicians be rated according to adherence to "Standards of Care," "quality assurance protocols," and "evidence-based medicine." These are basically protocols to be followed for treatment of all patients with a given condition. If physicians follow these protocols and can document that they have done so, they are deemed to be good doctors. If not—well, then a doctor is the equivalent of a 1958 Edsel.

Some years ago, I had a seventy-nine-year-old patient we will call Charlie. He was fairly healthy other than long-standing hypertension—high blood pressure. He had been on the same medication for twenty-five years, and his blood pressure had been completely stable during that time except for the occasions when I had attempted to change his medication. I had tried to change it because the one he had been taking was a first-generation anti-hypertensive and had some dangerous side effects.

After my second unsuccessful attempt to change his regimen, he posed a simple question. "Hey Doc. Why don't we just go back to my old medication? It was working pretty good for a long time."

"Charlie," I said, "it's got some bad side effects. Some studies just came out that strongly suggest anybody taking it should be changed to one of the newer ones."

Charlie screwed his eyes closed in concentration and then said, "Doc, I was on it for twenty-five years, and I never had a side effect. You did blood tests and EKGs and things—right? Why would I get a side effect now?"

289

Medical Metamorphosis

I opened and closed my mouth two or three times without saying anything. Charlie's commonsense argument was watertight. I had allowed my own common sense to be shackled by *protocol*. I put him back on his earlier medication, and his blood pressure settled back into an acceptable range. He never had any of the medication's dangerous side effects. Some people don't. *No two patients are exactly the same.*

Because I put Charlie back on his old medication, his insurer gave me a poor grade. I had not followed its protocol. Charlie never knew that, and I was glad, not because it would have damaged his trust in me, but had he known, he would have felt badly. It is circumstances like Charlie's blood pressure that make me skeptical of the suggested criteria used to grade physicians. As a patient, I'm pretty sure the fact that a doctor can follow cookbook instructions does little to fire up my trust that he or she care enough about me as a patient to consider my particular needs and lifestyle when suggesting plans of treatment.

I am not discounting the concept of standardization of care and the efforts to define appropriate baseline treatment protocols. There are doctors who need such parameters because for whatever reasons their ways of treating problems are haphazard. But systems, grades, and a medical equivalent to the USDA meat rating system cannot replace the personal trust created by having an ongoing relationship with your own doctor.

As a matter of fact—and I want you to look deep within

your heart and soul before answering—how many Americans are educated consumers in *any* area? I find it hard to believe that there are huge numbers of people who can make any more sense out of those ratings charts than I do. "Okay, let me see. This television got four stars for contrast but only three and a half for clarity while the other one got five stars for clarity but only three for contrast, but the first one got five for brightness while the second... let me see... no that's wrong. I got the columns mixed up. Those were actually the ratings for food processors. But wait a minute. What the hell does contrast have to do with cutting up vegetables? No, no, no. I'm losing it. I got to start over. Okay the first television can puree a picture into a bright piece of broccoli...."

In my personal life, I pursue personal trust whenever I can. I've used the same barber, gone to the same auto mechanic, taken my clothes to the same dry cleaner, been cared for by the same dentist, and used the same accountant for decades. I trust these professionals because they've always acted in my best interests. For me, being able to assume that someone *will* act in my best interests outweighs all other criteria.

A while ago, I took my car to my mechanic, Steve, because it was making a very unusual clunking sound. Steve climbed into the front seat and started punching buttons on my radio. A printout popped up where the radio station broadcast frequencies were normally displayed.

"I could have done that," I said. "What station do you

291

tune to?"

Steve smiled his all-suffering smile, started the car, climbed out, and opened the hood. He cocked his head to one side and listened intently. I could not help but think of a cardiologist listening to a heart. Then he plunged his head under the hood and after a few moments gestured for me to do the same.

"Look," he said, pointing to something that was bolted to something else, which had lots of wires coming from it. "Watch what happens when I crank up the engine."

He touched something else, and the engine began to rev up. As it did, the thing with the wires began to shake and the other thing, which was bolted to it, began to go "clunk -- balunk -- clunk." He stepped back from the engine and began writing on his work order. "For some reason," he said as he wrote, "they decided to use a real small bolt to secure that thing. I've seen it shear off before. Problem is, it's a tough area to get to."

I nodded, knowing he was gently implying it was not covered under warranty on a car with 120,000 miles on it, and that lots of labor would be involved.

"By the way," he added, "the all-knowing diagnostic program said everything actually was okay in spite of the sound, but that you should rotate your tires."

We both laughed. I did not ask him what might have happened had I ignored the clunking sound, because those are the sorts of things I don't want to know, much like patients who don't want to know what would have happened had they continued to ignore their chest pain. I was grateful that *my* mechanic had the "clinical acumen" and the experience with my model car to diagnose the problem before whatever would have happened, happened. I was also grateful when he looked at my tires and said they did not need to be rotated.

That's why I have followed my mechanic as he has moved from one dealership to another. It's not just his long-term relationship with my car that creates my loyalty. It's also the fact that I *trust* him. He could have rotated my tires, but he did not. I suppose that if I were an auto aficionado, I would have been able to determine whether the tires needed rotation, but I'm not. I will defer to my mechanic because I know that while he makes his living repairing cars, his ethics are equal to his skills.

I think it impossible to overestimate the value of being able to trust your doctor. The most efficient and logical way to achieve that trust is to create a relationship with a primary care doctor and guard that relationship and continuity of care as though your life depended on it. It might.

Trust and the cost of health care

After I left my practice, I did an informal survey of fifteen people. None of them were my patients. They varied in economic and educational status. The survey consisted of one question, "What do you want in a doctor?" Its lack of scientific design makes its results of no interest to anyone other than myself—and

293

now you.

Every one of the people I asked, one hundred percent, gave me an answer that centered on the concepts of trust and caring. My interviewees were universal in their desire to have a physician that they felt cared about them and whom they trusted. Price, education, board certification, and whether or not a doctor had been sued were never mentioned. Competence was mentioned eight out of fifteen times, and always as a second criterion. Of the fifteen, only two said they had doctors who met the criteria of trust and caring.

I've seen lots of studies about what patients want in a doctor. They are all much more detailed, better structured to eliminate sample error, exponentially larger than mine, and less conclusive. I'm sure you'll not be surprised when I tell you I believe mine more than the others. It is not just my iconoclastic bent that makes this so. It is also because the importance of these two concepts—trust and caring—was pounded into me over and over again during my thirty years of practice.

At present, a trusting doctor/patient relationship is rarely available in our health care system. Only two of the fifteen people I interviewed had one. All the factors associated with trust—continuity of care, established and stable PCPs with the time to listen, an insurance program that values stable doctor/patient relationships, and a patient oriented system—have been purged from health care as it was transformed into an industry.

The experts have assumed that patients will be able to trust in depersonalized health care the same way they trust in airlines. They propose that the abstract concepts of an "educated patient," "consumerism," and various forms of quality assurance replace the bond of trust between a patient and doctor. This is another example of abstraction ignoring reality. Flying to Buffalo is one thing. Dealing with your own mortality and human frailty is something altogether different. The issues involved in health care are far too intimate, too personal, and too powerful to lend themselves to impersonal, abstracted trust. Without trust, the specter of illness or injury can become even more stressful and frightening than it already intrinsically is. Fear is often expressed as anger or frustration, and these emotions can cause disruptive behavior. Without doctor/patient relationships characterized by trust, a health care system can be chaotic.

I want to put you in another pair of shoes. The shoes are size eight Brunomagli t-strap pumps with a two-inch heel. Your name is Gloria Heiden. You are thirty-one, almost divorced, living alone in a town house. You are an executive vice president at a midsized bank. This is your first appointment with Doctor Fitzgerald, a general internist. You have not had an ongoing relationship with a doctor partly because your health insurance has changed three times in the last five years and partly because of your own personal inclinations.

You are Gloria.

You are just about to walk out of the exam room.

You've been waiting in it for twenty minutes. On the exam table sits the gown the nurse told you to put on. You saw no reason to, but avoided what you thought would be a senseless confrontation by simply nodding and smiling.

You are slightly startled as the door suddenly swings open. A rather short man in a white lab coat walks in. He has thick gray hair and a tanned but ruddy face. He has reading glasses perched almost on the end of his nose. He looks at you over their tortoiseshell frames, smiles and says, "Ms. Heiden. I'm Doctor Fitzgerald. So nice to meet you. How can I help you today?"

He has not extended his hand, but you stand and extend yours. You notice that his hand is quite cold. You grip it in as firm a handshake as you can without trying to flaunt the strength you have from years of working out four times a week. "Well, doctor," you say through your most disarming smile, "I might tell you that the first thing you can do is to be on time." You broaden your smile into a soft laugh. "But I won't. Many of my best customers are doctors, and I know how busy you people are."

Doctor Fitzgerald is slightly taken aback. He looks at your chart. "You work at a bank. I see. What do you do there?"

"I'm an executive vice president in the department of business loans." You remain standing.

Doctor Fitzgerald walks to a rolling stool adjacent

to a small writing desk but also remains standing. "Do you have some specific medical problems Ms. Heiden, or is this a get-acquainted appointment?"

"Well, what I actually need are some tests."

"What sorts of tests?" Doctor Fitzgerald takes off his glasses and places them in his coat pocket. He looks at you expectantly.

From your purse, you pull a small notebook. "Ms. Heiden, wait," interrupts the doctor. "Before you tell me what tests you've decided you need, how about we discuss your history, your symptoms?"

You smile again. "Of course."

Doctor Fitzgerald sits down at the writing desk and begins asking you questions. The first questions concern your past medical history, drug allergies, and any ongoing medications. Your answers are honest but brief. When he finally begins asking about the specific reasons for your appointment, you are more hesitant. Your symptoms have raised specific concerns, and you are afraid that if you fail to appropriately describe those symptoms, the concerns will not be addressed. Carefully, you say, "Well, the greatest problem I'm having is with headaches. They have become progressively worse over the past few months. Now they are almost daily, and to be quite honest, Doctor Fitzgerald, they have started to interfere with my ability to do my job. I would really like to see a neurologist."

Doctor Fitzgerald nods his head but says nothing. You are about to volunteer more history, but instead say, "I'm quite sure that the neurologist will want an MRI, and I would like to go ahead and schedule one of those prior to my visit."

The doctor smiles but still says nothing.

"And I would like to get some lab tests done." You find yourself talking quickly. "I think the neurologist will want lab tests, and it's been some time since I had my cholesterol checked." You tear off a sheet from the notebook. On it is a list of thirty blood tests. You thrust the sheet towards the doctor

Doctor Fitzgerald looks over the list and smiles. Your face flushes slightly. You prepare yourself for the argument you've been dreading, the argument about keeping down costs and unnecessary testing, the argument that is also about control of your own health care. He looks up and says, "I'm just guessing, but are you worried about diseases such as multiple sclerosis and lupus?"

"Well ... a bit. They are common causes of my symptoms."

"What about tumor?" he asks. "Does that worry you?"

You look at his face, trying to see any sign of sarcasm. There is none. "Not much."

He nods. "Good. That's not likely. Neither is lupus or M.S."

"How can you possibly say that, doctor? You have not even examined me? How can you expect me to...?"

He cuts you short. "Ms. Heiden, don't worry. I'm going to send you to the neurologist, order your MRI and the entire list of lab tests, plus a couple other tests that you need. You have a gynecologist don't you?"

"Of course."

"Good." He pulls a lab request from a drawer and fills it out.

"You'll make sure I get a copy of the lab?"

"Of course." He hands the request form to you, and stands. He takes a step towards the door.

"Will you go over it with me? Should I see you after I see the neurologist?"

He stops. He takes off his reading glasses that he had put back on while filling out the lab request. Instead of putting them in his pocket, he appears to study them. Finally, he looks up. "Ms. Heiden, believe it or not, I *am* bothered by running late, so I should probably just say yes and leave. But that bothers me also.

"You see, I'm a short-timer in this job. In a few months, I will retire from medicine. I work this job because I sold my practice to a hospital, and part of that sale involved me continuing to work for a period of time.

"Now, those for whom I work want me to see patients every ten minutes. I'm not sure what I'm supposed to do in ten minutes, but it's not to practice medicine. The only way I have been able to do what I am supposed to do is to work very hard at not caring about those I see. But that's not easy.

"I suspect that your headaches are muscular in origin. I suspect they are present because something has you wound up tighter than a cheap watch. You have been clenching your teeth the entire time we have been talking."

Doctor Fitzgerald walks over and reaches towards your face with his right hand. "May I examine your jaw joint?"

You nod. He places his index finger just in front and slightly below your right ear. As he applies pressure, you wince. "You're cranking on that as hard as you can," he says. "Hasn't your dentist said anything about your grinding of teeth?"

You smile sheepishly. "Well, it's been awhile since I've seen my dentist."

He laughs. "Shame, shame, shame. You probably don't floss either. Well, I suggest you see him. A bite-block might help your clenching."

He puts away his glasses. "You may also want to spend some time looking at what might be *causing* you to clench. As for the lab results, you'll get a brochure that explains what they mean."

He draws in a long breath. "Look, for what it's worth, the chances of these headaches being life threatening are less than the chances of you having an accident driving home. Go ahead with your tests, but try not to worry about M.S. or a brain tumor."

He quickly leaves the room. As you walk to the checkout station, you unconsciously rub your jaw. While waiting in line, you jot yourself a note to investigate teeth clenching on the Internet when you get home after work.

Then, you remember you have an early evening appointment with your divorce lawyer. Your forehead furrows and the muscles of your jaw ripple as you react to that recollection.

Odds are substantially in favor of Gloria's MRI being normal. Odds are substantially against her headaches improving, even after seeing the neurologist. The most likely cause of her symptoms is bruxism (teeth grinding) and secondary muscular contraction headaches. This is a problem that involves the dynamics of the jaw joint, the muscles of the head and neck, sometimes a patient's work environment (long hours at a computer is a common precipitant, particularly if a patient's workstation is configured poorly.), and often the patient's emotional state. The inherent complexity of the interaction between these factors demands a very integrative therapeutic approach. Few neurologists have the time to invest in such an approach.

Gloria would appear to be the epitome of an "educated patient." She is bright, motivated to take control of her own care, and has access to unlimited medical information via the Internet. But she will probably continue to suffer from headaches because they are the result of a complex, multifaceted process. There is no simple solution to such problems. It has been my experience that their improvement requires the effective application of holism, education, pragmatic use of multiple treatment modes, and the art of healing. These are almost always centered on a consistent and caring doctor/patient relationship.

Hopefully, Gloria will not embark upon what I call the *educated patient's pilgrimage.* This is a self-directed journey from doctor to doctor, treatment to treatment, diagnosis to diagnosis. It is a journey accompanied by the patient's progressive deterioration not only because the primary problem is never addressed, but also because a patient can only undergo so many treatments before encountering side effects, either from the treatments themselves or interactions between them. It is a journey that may sometimes carry the patient to a variety of different forms of alternative medicine, many of which are often quite expensive. An educated patient's pilgrimage is always initiated by lack of trust within the doctor/patient relationship or no doctor/patient relationship at all.

In my thirty years of practice, I saw hundreds of patients

who had taken the educated patient's pilgrimage, many because of this type of headache. Some of them were on multiple medications. Some had become addicted to pain medications. Sometimes, I was able to help these patients, sometimes not. Establishing a relationship of trust was difficult because they had learned *not* to trust either the system or doctors. Both had failed them. In addition, these patients required a large investment of time, patience, and, above all else, persistence. I was not always able to successfully make this investment.

In Gloria's case, we have a bit of a clue about one of the factors causing her headaches: her impending divorce. An acutely stressful circumstance is often associated with bruxism and headaches. She has also not seen her dentist in some time. If she has a predisposing dental condition, bruxism is a common consequence.

If Doctor Fitzgerald had been able to establish a relationship of trust, these factors might have been elucidated. That does not mean they could have easily been mitigated or even that Gloria would have accepted their role in her headaches. Often, a patient would rather worry about having a debilitating disease like M.S. than openly confront the possibility that lifestyle or relationships are playing a role in their symptoms.

Doctor Fitzgerald made no attempt to earn Gloria's trust. It's apparent he feels guilty about not doing so because he makes a rather surprising confession that he is a "short-timer." It's also apparent that he's just punching the clock until he can retire.

303

Because he can no longer fight a system that minimizes the importance of the doctor/patient relationship, he acquiesces to expensive tests and specialty referrals, even though his clinical instincts suggest these are not needed and will not help. Doctor Fitzgerald's clinical skills have become irrelevant in the context of an educated patient and a medical *industry*.

If I had an established relationship with a patient, their demand "I want an MRI," if inappropriate, could often be defrayed with a discussion of why it was not needed. If I had not established a bond of trust with a patient making such a request, discussion was futile. The irony of this circumstance is that the system criticizes practitioners for practicing defensive medicine while discounting the trust created by an ongoing doctor/patient relationship. They also criticize practitioners for overutilizing technology. It is not just defensive medicine that creates overutilization. Nor is it an obsession with technology. It is a disruption of the relationship between doctor and patient. It is the relentless restraint of clinical judgment as an appropriate alternative to expensive procedures.

When Governor Lamm suggests that the doctor/patient relationship should not be an integral part of the economics of health care, he dismisses the only factor that might decrease overutilization. The erosion of primary care and the disappearance of healers have left patients with no one to trust but themselves. If you had a severe headache and ended up in a busy emergency room, were seen by a doctor you did not know, spent only a very brief period of time with him and were told you probably just had a stress headache, would you demand an MRI? An "educated patient" would. The experts are suggesting everyone should be an educated patient. Guess what? Now the emergency room physicians don't even wait for patients to ask. They do MRIs on everyone with a headache. I don't blame the patient. Nor do I blame the emergency room physician.

When those factors that nurture trust between patient and doctor are subverted, the only remaining part of the "miracle of modern medicine" is technology. That the economists and academicians are surprised when technology is overutilized is simply tribute to how clueless they really are.

Medical Metamorphosis

Truth, like light, is blinding. Lies, on the other hand, are a beautiful dusk, which enhances the value of each object.

-- Albert Camus in *The Fall*, p. 126, Gallimard (1956).

A Short History of Medicine

2000 B.C. - "Here, eat this root."

1000 B.C. - "That root is heathen, say this prayer."

1850 A.D. - "That prayer is superstition, drink this potion."

1940 A.D. - "That potion is snake oil, swallow this pill."

1985 A.D. - "That pill is ineffective, take this antibiotic."

2000 A.D. - "That antibiotic is artificial. Here, eat this root."

--Author Unknown

CHAPTER EIGHTEEN

THE FOG

Melodrama

Sources of information about health care have sprouted on the Internet and cable networks like mushrooms after an all-night rain. The problem is that like mushrooms, many of these sources of information can be poisonous. Hucksterism, radical philosophies, and elemental stupidity abound but often look too delectable to resist. It's difficult to ignore a convincing argument when it promises painless weight loss for only $39.95 a month. And what aging lothario could pass up the opportunity for the virility of a stallion when it can be ordered anonymously on the Internet? Of all these mushrooms of information, however, melodrama may have the most classic flavor.

Melodrama is a dramatic style based upon simple concepts, simple ethics, and simple characters. It reached its zenith more than one hundred years ago but has survived the ensuing century, demonstrating that obvious heroes, obvious villains, and an uncomplicated depiction of good and evil have an unending appeal.

Two of the more popular examples of contemporary melodrama are professional wrestling and American partisan politics. On rare occasions these two melodramatic forms even intersect. Jesse Ventura, a professional wrestler, served as the governor of the state of Minnesota. Conversely, Bill O'Reilly, a political critic, aggressively pursued an avocation of amateur wrestling, but his dreams were limited to unsuccessful telephone negotiations. We are blessed that such intersections are rare. It is painful to contemplate the possibility of Rush Limbaugh, Al Franken, or Michael Moore playing their melodramatic roles while wearing Speedo wrestling briefs.

While the essence of melodrama is simplicity, both wrestling and politics have successfully interlaced basic elements of good and evil into sophisticated forms of entertainment.

Professional wrestling utilizes scripted scenarios modeled after soap operas. Clashes between good and evil are enhanced by sex, alliances between individual wrestlers, and vendettas. Characters sometimes also change roles. Hulk Hogan, for example, was at one time a villain, then became a hero, and is now a villain (when last I checked).

Political melodrama, on the other hand, allows combatants to concomitantly identify themselves as heroes *and* villains. Liberals and conservatives, Republicans and Democrats, *all* participants in political melodrama are simultaneously both hero to some and villain to others. Rush Limbaugh does some ranting, and half the public is enraged while the other half is screaming "You tell'um Rush! I've been a ditto-head for five years." Limbaugh twirls his mustache and rides a white horse at the same time. Thus, the point-counterpoint of good and evil does not need a hero and a villain on stage at the same time because a single actor plays both parts simultaneously. It's economic as well as effective melodrama, guaranteed to engender an emotional response in the maximum percentage of an audience at any given point in time.

The actors in political melodrama are emotional provocateurs as skilled as any snarling, face-painted wrestler or his heroic all-American adversary. Rush Limbaugh's pomposity—"Rush Limbaugh... here as a gift from God..."—is every bit as effective as a body slam and knee to the Adam's apple. It's the stuff of well-constructed melodrama.

The problem is that this melodrama takes a huge toll on truth and objectivity.   Go to a movie by Michael Moore and then race home and watch the "O'Reilly Factor." If you aren't just a bit confused, then you believed one of them and considered the other an absolute liar. There's not much middle ground because when they discuss the same issue, they come to exactly opposite conclusions. You have been forced to *believe* one or the other. The opportunity for discerning thought has been lost.

Let's be honest—neither side wants you thinking. You're supposed to be cheering. That's what you do when watching melodrama, you cheer. It's not supposed to make you think. Boo to the villain! Hurrah to the hero! Boo to The Crusher! Hurrah to The Rock! Bill O'Reilly is a misogynist. I believe Al Franken! I believe Bill O'Reilly. Al Franken is a dork!

The operative word in all of this is "believe." Whom do you believe? Who is telling you the truth? Is America about to dissolve financially because of its budget deficit or is the deficit really not all that important? Are the present efforts to protect America from acts of terrorism effective or are they exercises in futility that have squandered billions upon billions of dollars and thousands of lives? Rush Limbaugh will give you one answer. Al Franken will give you its opposite. Whom do you believe when those offering opinions are playing melodramatic roles of exaggerated simplicity?

Crises of Contrivance

It would be nice if truth about medical care were less elusive. The amount of knowledge generated by medical research is enormous, and a dependable interpretation of this knowledge would be a valuable resource. Don't hold your breath while waiting for reasonable interpretation in the media or on the Internet. You'll turn blue. The politics of health care produce some of America's most colorful melodrama, and the science itself is rife with conflicting advice and opinions, contradictory warnings and admonitions, and general confusion.

This is another reason that primary care must become health care's strongest specialty. One of a PCP's roles is to help you wade through those melodramatic chunks of information stuck in your face while watching the morning news.

If you have your own doctor, you have a trusted partner with whom you can discuss your concern after you see a smiling anchor-person hold up a bottle of your blood pressure medication and say, "Well, if this happens to be a medication you're taking, you may be interested in a report out of Okefanokee University. Basically, the report says ...YOU'RE GOING TO DIE! Heh, heh. Don't buy any long playing records, my friends."

Should you worry? Should you withdraw your entire IRA and buy that sports car you've lusted after? Might as well if for you there really *is* no tomorrow. Or should you write a letter to the producer of *Early Morning Fast Action News* and tell him his anchor-person is a jerk because he needlessly ruined a perfectly

311

good bowl of Trix.

We are besieged by what I call *crises of contrivance.* These are medical issues, usually controversial, that become raised to the status of life-threatening catastrophes. They are elevated to an attention-grabbing status because someone needs for them to be attention grabbing. They are created to serve an economic end, a particular dogma, or the pursuit for fame and power. The media enthusiastically adds to the melodrama and gravity of these crises because melodrama captures attention.

If evaluated objectively and without a self-serving agenda, these crises are rarely the disasters as portended. *Crises of contrivance* are a form of melodrama. They are subtler than professional wrestling and talk radio, but they are still exaggeration for effect.

Estrogen use, generally referred to as hormone replacement therapy (HRT), is a good example of a *crisis of contrivance.* For almost four decades, HRT has been the pig-tailed little girl of the pharmacological world. She has been called very, very good—saving many women from the incapacitating physiology of menopause, thought to be cardioprotective, thought to protect against cancer of the colon, and even described as perhaps prolonging mental acuity. She has also been castigated as being very, very bad—described as perhaps playing a causal role in cancer of the breast, and instead of protecting the heart, actively causing cardiac disease resulting in sudden death. Most recently, a large study placed her in the latter category—very,

very bad.

The Women's Health Initiative (WHI) was the largest study of the effects of HRT in post-menopausal women. The study was stopped well before its projected completion because of concerns at the National Institute of Health (NIH) regarding risks to the study's participants. Those participants taking the estrogen/progestin combination were having an increased incidence of cardiac and embolic side effects. This was a surprise to most authorities in the area of HRT because they had believed that the pigtailed little girl was very, very good. Until the WHI study, it was thought that HRT actually *prevented* heart disease.

What a shock. What a crisis. What controversy. What a way to fill those science spots on the cable news networks. It was overtime for the TV doctors with big hair. The trial lawyers wanted to declare a national holiday celebrating the potential of billions of dollars of lawsuits.

Unless you were a participant in the study (16,000 women were involved, and they were sent letters before the general notification), the announcement of the study's cessation came in the form of a drama-filled press conference. As I remember, there were four rows of microphones arrayed in front of the lectern. Four rows of microphones indicates that *someone* thinks what's about to be said should be heard by lots and lots of people. A spokesman slowly approached the lectern and paused. He was wearing a long, white lab coat that didn't quite fit. The room went silent, and with great solemnity, he began his fifteen minutes of Warholian fame. "We're announcing something that's *never* been done. Because of risks to its participants, we are *completely* stopping a major study, the *largest* study ever done on hormone replacement therapy... "

I had two reactions to the announcement. The first was a desire to roll around on the floor and cry. These sorts of announcements are the equivalent of not only shouting "FIRE!" in a crowded theater, but adding "AND I'LL BET ALL THE DOORS ARE LOCKED!"

Typically, these sorts of announcements begin by cataloging all the possible consequences of some treatment, medication, or disease in language that sounds vaguely like the climax of a short story written by Edgar Allen Poe. There is then a momentary pause to let a complete understanding penetrate the areas of the brain associated with panic. Then the disclaimer is made that if any part of this tale of horrors has made you the slightest bit uneasy, you should immediately call your personal physician. Granted, it may be eight o'clock on a Saturday night, granted no one has had the courtesy to warn your personal physician of this dramatic announcement since it has been made to the national media *before* appearing in the professional literature, and granted you may have been on the medication for ten years and the chances that something awful will happen in the next twelve hours are pretty slim. Nevertheless, if you're even a tad worried, call your doctor so you can become angry and accuse him or her of being unprofessional when they honestly admit they have no idea what the announcement really means. The reason I did not roll around on the floor on this occasion was because I was not on call. I did, however, light a candle for my partner who was.

My second feeling upon hearing the announcement was a vague sense that something was not quite right. Something did not make sense, but I wasn't sure what.

My very first patient the next day defined the what. She was a postmenopausal woman who had been on estrogen replacement for eleven years. She also had high blood pressure. Making things more complicated was the fact she had a strong family history of colon cancer. (The WHI study did verify a finding noted in earlier studies that hormone replacement therapy reduces the chances for developing colon cancer.) I reeled at the thought of trying to sort through a study that implied she should stay on the HRT for cancer prevention but stop it because she had high blood pressure. My patient's appointment was just for fifteen minutes, to check her blood pressure treatment. I already had six phone calls to return concerning the WHI study. Rolling around on the floor looked appealing.

Fortunately, my patient was a retired biologist blessed with an abundance of common sense. I started into a discussion of her hormones and she waved me off. "Forget it," she said. "Swimmers, non-swimmers."

I immediately knew what she meant. My vague uneasiness of the prior evening was replaced by an anger that

315

would grow for months. People are different. Making a broad sweeping statement about something like HRT is akin to lining up a hundred people at the edge of a huge swimming pool, asking them to jump in, and when ten of them drown, drawing the conclusion that swimming pools are *incredibly* dangerous. A ten per cent mortality rate is completely unacceptable -- "EVERYBODY OUT OF THE POOL!" But wait—shouldn't we know how many of those one hundred people were swimmers? Shouldn't we know if some of those people were wearing lead-weighted combat boots? That's more important than the depth of the water.

With regard to the swimming pool of hormone replacement therapy, we know *for a fact* that some women are swimmers and some women are wearing those lead-weighted combat boots. There are some women who have what is called a factor five mutation. This is a genetic condition present in about 3% of the population that can substantially increase the incidence of blood clots. It is a circumstance that is *definitely* worsened by estrogen therapy. Women with a factor five mutation are non-swimmers who are also wearing lead boots—hormone replacement therapy for them can be deadly. It can lead to the exact cardiac complications seen in the study. But the women in the WHI study were not screened for this genetic disease. So theoretically 3% of them had a factor five mutation, and we already know they should *never* have been placed on HRT in the first place.

On the other hand, there was a 37% *decrease* in the predicted cases of colon cancer in those who were on HRT. The population of participants who were at greater familial risk for colon cancer were not studied separately, so it must be assumed that the reduction of colon cancer in this specific population, as opposed to the general population, was even much higher.

Women with a familial risk for cancer are not only swimmers, but jumping in the pool of hormone replacement therapy may *save their lives*. With her family history of colon cancer and having been on HRT for years without cardiac problems, my patient knew she was a swimmer. She had no intention of getting out of the pool.

The WHI study is a classic *crisis of contrivance.* When looking at the data, if 10,000 women were placed on HRT, the predicted number of heart attacks in women on estrogen plus progestin is 37 compared to 30 on placebo. That's an extra 7 occurrences or 0.07 %—*less than one occurrence per thousand.* The predicted increase in number of strokes is 0.08 %—*less than one occurrence per thousand.*[5] And these increases would assuredly have been even lower if women with factor five mutations had been screened from the study. These are very small increased risks, not even in the same ballpark as other risky circumstances that experts totally ignore—like not having health insurance. The risk of not having health insurance in America is

---

[5] As reported at the website of the Woman's Health Initiative, http://www.whi.org

the same as a person being a brittle diabetic, substantially greater than taking HRT.

Are there women who should *not* be on HRT? Yes. Are there women who *should* be on HRT? Yes. If concern is about patients, this is the question—where does a patient fit into the scheme, swimmer or non-swimmer?

But that was not the concern. Following that press conference, rhetoric and agendas, controversies and recriminations, and ads for lawyers specializing in HRT all mushroomed into another field of confusing information and claims. Women's groups accused "the male-dominated medical industry" of "medicalizing" the natural process of menopause. The pharmaceutical company that manufactured the specific medication used in the study was excoriated as being no better than tobacco companies. At the same time, two other manufacturers of alternative hormonal medications ran ad campaigns stressing that their products had not been implicated in the study, ignoring the lack of any solid evidence that their side effect experience would have been different. Economists expounded upon the cost of unnecessary medication. Investigative reporters hinted at the role of influence peddling involving the pharmaceutical industry and physicians. And the patients were forgotten. Most women were summarily either taken off HRT or changed to one of the "safer ones." But "safer" was defined more by legal concerns than by any honest physiological criteria.

As contrivance and hypocrisy fueled this particular crisis, I think a few of my associates thought I had finally lost the last of my marbles. I ranted about swimmers and non-swimmers. I pointed out that treating all women the same with regard to HRT was like ignoring the difference between a sore throat in a hippopotamus and a giraffe. (I don't think a hippo even has a throat, and a giraffe is nothing but throat.) Every once in a while I managed to produce a mumbled agreement, but usually my consternation was countered with a warning. "Listen, Doctor Waggoner. If your patient has a side effect, even if it was in somebody who probably should have been on HRT, your goose is cooked."

Therein lays the greatest risk of *crises of contrivance.* They obscure the truth. Rhetoric and self-serving agendas can so embellish sophistry that rational thought and reason have no voice in matters of significance. The music of melodrama drowns out common sense.

I am not suggesting that concerns about America's health care system are all *crises of contrivance.* However, an honest assessment of an American health care crisis must include a sense of perspective. There are a variety of interests who will extrapolate issues beyond the realm of reasonable perspective, and regardless of whether their agenda is well meaning, self serving, or purely economic, the net result is still confusion and acrimony.

It is the average American, the patient, who is smack in

the middle of the confusion. On Monday, hormone replacement therapy protects women from having premature heart disease. On Wednesday not only does it not protect against heart disease, it *causes* heart attacks and strokes. On Wednesday, Vioxx is a medication that eases the pain and inflammation of arthritis and is also safer than the older nonsteroidal anti-inflammatories because it does not cause gastrointestinal bleeding. On Friday, even being in the same room when a bottle of Vioxx is opened can make a person instantly prone to strokes and heart attacks.

And over the weekend, the hay fever medication you had been taking for decades is removed from the market, apparently capable of causing fatal arrhythmias. You wonder a bit why the medication is now a risk given the fact you've taken it with every ragweed bloom for thirty years. Why is it suddenly risky? Are there other factors at work? How is it possible that medical opinion can change so dramatically?

A few months later, you see that Vioxx's parent pharmaceutical company loses its first liability lawsuit. This is a bit surprising since the patient in whose name the lawsuit was filed died because of an irregular heart rhythm. You've never heard that an irregular heart rhythm is one of Vioxx's side effects. The mystery is explained when you read that even though an autopsy on the patient did not show evidence of a heart attack, an expert witness hired by the plaintiff testified that the patient probably had one anyway. You wonder a bit about the reasons for doing an autopsy if its results are simply tossed out the window.

Then, you read that an expert from a large prestigious medical school has offered the opinion that estrogen is not as bad as was suggested at the conclusion of the WHI study. You are still scratching your head about that opinion, given the absolute revulsion surrounding estrogen only months earlier, when you hear that just like Vioxx, the older nonsteroidal arthritis medications can cause heart attacks. Even the old standards like ibuprofen and naproxen, now sold across the counter, are suspect. You hear a doctor on television talk about learning to live with pain and how reducing certain kinds of activities will reduce inflammation.

It occurs to you that modern medicine has taken a thirty-year step backwards. However, you next hear an arthritis specialist say that since Vioxx isn't really a greater risk than older medications with regard to causing heart attacks, and since the older medications cause significantly more gastrointestinal bleeding, it probably makes sense to put Vioxx back on the market. You hear this opinion the same day you hear that the next Vioxx lawsuit has started, and that it would appear tens of billions of dollars are at stake. The court reporter adds that behind the scenes, experts are saying there is no way Vioxx could have played a significant role in the first case, and it would appear juries want to "sock it to" pharmaceutical companies.

Meanwhile, FDA officials start slugging it out—with each other. Accusations fly fast and furious, running the gamut from *rushing* drugs onto the market because of influence applied by pharmaceutical companies to *delaying* the release of drugs because of political pressures. You are not particularly surprised when the head of the FDA calls it quits. Under the circumstances, it seems like a sane decision.

The melodrama accelerates to a climax worthy of *A Night at the Opera*. The Marx Brothers, however, made much more sense than the medical authorities. You are completely confused. You wait for a duck to drop down with an explanation, but there is no duck. A scientific understanding of Vioxx, or estrogen, or the decongestant in an allergy pill has nothing to do with these swings in opinion. They are *crises of contrivance*.

For example, the decongestant in your allergy pill, phenylpropanolamine, is a medication that has been used in hundreds of cold and allergy preparations. When it was removed from the market, it had been taken in billions of doses. As it was pulled from the shelves of pharmacies and grocery stores, there was still *no* study that effectively linked it to a significant increase in hemorrhagic strokes in women—the theoretical reason it was taken off the market.

It was not a piece of research that removed this medication from the market. It was the loss of a 1.5-million-dollar lawsuit, the subsequent posturing of "blow the whistle" academicians, and a feeding frenzy of trial lawyers. The drug's manufacturers saw a perfect storm of liability on the horizon and decided to beach phenylpropanolamine as quickly as possible.

Did phenylpropanolamine have side effects? Absolutely.

Anything people place in their mouths, including organically grown, herbicidal-free, handpicked vegetables, has the potential to wreak havoc on their bodies. Ask a person who has had her first allergic reaction to a peanut whether this benign appearing member of the pea family can cause problems. It is estimated that between 50 and 100 Americans die annually because of peanut allergies—far more deaths than were ever associated with phenylpropanolamine. So much for peanut butter—rip it off the shelves.

I prescribed countless doses of phenylpropanolamine, but I did not use it where it was likely to cause more harm than good. That is simple common sense. Most of the documented cases of significant problems with this drug were either in a setting of overdose or use in patients who should not have been on it in the first place. I did not recommend it to my eighty-year-old patients who had high blood pressure. Nor did I suggest that any patient take five times the recommended dose. I also didn't give Reese's Pieces as treats to my pediatric patients who had peanut allergies.

Unfortunately, the melodrama of *crises of contrivance* is impacting the lives of Americans just like the melodrama of partisan politics. Political figures that speak of compromise are scorned for being disloyal to their parties. Diplomacy is no longer a virtue. The pundits categorize medications as being good or bad rather than appropriate in some circumstances and not in others. Clinical judgment and common sense have been sacrificed on the altar of so-called "evidence-based regimens." The legislative process is a shambles, and in health care, truth, common sense, and reason have been locked in the basement.

So whom *do* you believe? The only trained professional capable of wading through this muddy water and offering reasonable advice about *your own* health care is *your own* doctor. He or she is the only medical professional who knows the details of not only your physiology but also how those details fit into the rest of your life. *No two patients are exactly the same.* Regardless of claims and counterclaims, this simple observation is the cornerstone of any truthful conclusion to a medical controversy.

Genomic medicine is matching treatment to a patient's genome. Some clear-headed medical authorities—and believe it or not, they do exist—think this may revolutionize medical care. But even now, in the early stages of this new discipline, it is recognition that *no two patients are exactly the same.* Already, cancer drugs that had been abandoned as being ineffective have proved to be life-saving for a specific population of patients. Decisions regarding what may or not may be appropriate for you must be made with *you* in mind—not the melodrama of medical controversy. The Internet may provide wonderful access to this controversy, but it's virtually devoid of patient-specific information. The doctor with the big hair may offer advice about a problem that is appropriate for ninety percent of the population, but if you're a member of the other ten percent, it's inappropriate for you. Whatever you have, you have one hundred percent of the time. That makes the other statistics rather irrelevant. You need your own doctor to discuss the population that determines what is best for you—that population of one, a population comprised of you and only you.

Snake oil

In the American west of the late 1880s, traveling melodrama shows were often accompanied by *snake oil salesmen*. They sold elixirs and remedies that were touted as being effective for everything from lumbago to the common cold. Often the active ingredient in these medicinal preparations was alcohol, opiates, or cocaine. They might not have cured much, but they were understandably quite popular. Americans now spend billions of dollars on what is called *Alternative Medicine*. This concept has even become a medical discipline, which is why it is capitalized in the last sentence. Quite honestly, much of Alternative Medicine is no different from snake oil sales. What has changed to elevate a sideshow into a medical discipline?

What has changed is the purchase of respectability with those billions of dollars. What has also changed is Alternative Medicine's association with the valid concepts of *holism* and *holistic medicine*. These concepts were originally a response to traditional medicine's transformation into a disease-oriented discipline following the Second World War. Modes of therapy outside traditional medicine such as massage, manipulation, and therapeutic diets were characterized as being of very doubtful legitimacy. But as we've already discussed, as traditional
325

medicine became less personal, patients sought forms of therapy that acknowledged their humanity. They sought healers. The discredited modes of therapy capitalized on this and began emphasizing self-healing, health promotion, and holism.

Unfortunately, this philosophy was often less about humanism and more about the specific modalities and promotion of their practitioners. Alternative Medicine became a circus of weight loss medications, new age health cults, incredibly expensive vitamin regimens (in reality no more than cleverly packaged, run of the mill vitamins sold for 100 to 1,000 times their actual cost), jewelry, juicers, pyramids, patches, male enhancers, magnetic beds, and books that made claims running the gamut from the secret to living to one hundred-fifty as well as a way to enlarge breasts from an A-minus cup to a quadruple-D through the use of guava juice mixed with aphid dew.

This circus is a dangerous place to visit. It's definitely a risk to the average patient's finances, but it can also be a threat to his or her health. Alternative "dietary supplements" are not monitored by any agency or scientific body. Their content is often unknown. Their advertised benefits are by and large unproven and more often than not simply the creation of a clever entrepreneur capitalizing on the age-old penchant of people for buying things they know won't work while still hoping that magic will prove them wrong.

The magic of alternative treatments is widely defended by the entrepreneurs and those who believe in their products. This defense is every bit as aggressive as those mounted in the arena of partisan politics. Once again, the average person is forced to believe in one side or the other.

"Natural" medications, genuine healers

I want to again put you in someone else's shoes. They are size nine, navy blue pumps, although size nine-and-a-half would probably fit better. Your name is Bernice Mitchell. You were born in a small farming community forty miles outside of Paris, You came to America as a war bride. You have three daughters, and your husband is still living. His health is failing rather quickly.

Your own health is actually quite robust, although from time to time you have had periods of anxiety. These episodes sometimes lasted as long as four or five months. During those times, you were convinced you had a life-threatening illness even though many medical tests were performed, and they were all normal. When a doctor first raised the possibility that anxiety was the cause of these episodes, you were incensed. Your green eyes locked with the doctor's, and you said, "My dear doctor, this is *not* in my head, so you had better look elsewhere." He did.

You have been with your present doctor for more than a decade. He was the one who finally convinced you that anxiety was indeed the cause of your "spells." While you still believe that to be so, you are unable to recall how Doc, as he's called, explained their origin. You know it had something to do with
327

your husband, and the war, and your dead parents and sister, and coming to America. But when you try to put the pieces together, it's as though your mind slows and finally stops, like a car running out of gas as it goes up a very long hill. You must apply the parking brake so the car does not roll backwards.

Today you are seeing Doc to check your progress with your high blood pressure. You are Bernice.

The door opens and Doc walks in. He stops and smiles at you. He has always been absolutely professional, but a part of you knows his smile acknowledges your efforts to look nice. "Bernice. How are you? Clarisse isn't with you today?"

You smile back. "No, Doc. She had to work. As a matter of fact, I came here from work myself."

Doc's face screws into a look that asks a question. "I guess I thought you knew," you say. "I am working cosmetics again, just part time. They needed the help."

Doc nods. "I trust you're doing okay with the blood pressure medication?" He looks at the blood pressure taken by the nurse. "Ouch! One-fifty over ninety-five. That's no good." He rechecks your blood pressure. "Still up."

He sits down on his rolling stool and looks at your chart. Finally he looks up. "You're not having any trouble with the medication?" He pauses for a moment, and then adds, "You're taking it, not missing many doses?"

Your face reddens.

"Bernice," he says, "you're not taking the medication at all, are you?" He stands and wags his finger at you.

"That's not so!" You do your best to sound indignant. "I *am* taking blood pressure medication. It's just ... not the medication you gave me."

Doc draws in a sigh while shaking his head. "Let me guess, Clarisse?"

You shrug.

He shrugs back. You reach into your purse, pull out the bottle of alfalfa extract capsules you've been taking instead of his blood pressure pills. You place them on the exam table. He picks up the bottle and studies its label.

You think for a moment, then say, "She looked up your medicine on the Internet. Doc, it has many, many side effects. We talked about it, and you know she works for the nutrition store. By the way, she is now an assistant manager. Anyway, she ... I mean *we* thought using a natural blood pressure medication would be better for me. Doc, I am just a simple French peasant girl. I don't need an expensive pill."

Doc rolls his eyes. "Bernice that line stopped working about eight years ago." You both laugh. "We talked about the side effects, Bernice. I gave you that entire list and wrote one of my own listing the problems I thought might be the most worrisome. Remember?"

He looks at you expectantly, and you again shrug. "Clarisse was very concerned. She truly believes that natural things are best."

Doc is quiet for a moment. Then he says, "Bernice, I know you grow vegetables in your garden. When you pull a carrot from the ground, do you typically wash it off, or eat it still covered with dirt? Do you peel it?"

Without thinking, you answer, "Don't be silly. I wash and peel it."

"Why wash it off?" he asks. "Isn't it more natural to eat it exactly as it came from the earth? And peeling it—doesn't that make it even more unnatural?"

You shake your head, but he continues, "What modern medicine does is wash off the carrot. The vast majority of prescription drugs were originally discovered in nature. But Mother Nature did not design our environment just for our own use. Most of the beneficial effects of what we find in nature come attached with unpleasant side effects. They are like the carrot, unwashed and not yet peeled. Modern medicine does to the medications what you do to the carrot. It washes and peels away what is not needed, what is harmful. And if you think pharmaceutical drugs have side effects, imagine what the venom of a Brazilian viper will do."

You do not like snakes and instinctively shudder. "The Brazilian viper?"

"The Brazilian viper. One of the most deadly snakes there is."

You make a face. "I hate snakes. What does a viper have to do with anything?"

Doc leans towards you just a bit. He opens his eyes comic-book wide and in a voice overly dramatic says, "It is the venom of the Brazilian viper that I gave you to control your blood pressure."

In spite of your best efforts your mouth still drops. "No."

"Yup." Doc leans back. "I gave you an ace inhibitor. Let me tell you how ace inhibitors came to be.

"Some years ago, an observant and quite clever scientist noticed that the Brazilian viper kills by dropping its victim's blood pressure down to nothing. All the blood vessels dilate completely wide open, the prey's blood pressure suddenly drops, and kaboom—the viper's victim bites the dust. It occurred to the scientist that perhaps the viper's venom might somehow be used to treat high blood pressure.

"I imagine he dismissed the idea of training vipers to deliver very small, non-fatal bites because Brazilian vipers have the reputation of being less than friendly. They would also be difficult to package in a bottle. So he decided to wash the dirt off the carrot, to analyze the venom and isolate the substance that lowered blood pressure."

Doc pauses and looks at you. "You with me?" You nod. He continues. "Doing this is a difficult chemical process, so the scientist took his idea—and I would imagine a bit of venom—to a pharmaceutical firm. For obvious reasons I doubt he took the snake. Two other scientists eventually synthesized the part of the venom that dropped blood pressure. They cleaned the carrot, that is, isolated the substance, and then 'peeled away' its bad parts, slightly changed the structure, such that it dropped blood pressure—but not all the way down to zero like the viper's venom did.

"In my mind," he says, finishing with a flourish, "that blood pressure medication is every bit as natural as any root or herb or alfalfa preparation that's supposed to reduce blood pressure. For heaven sakes, it's modeled after the venom of a snake. And it's about a million times more effective."

"Is all that true?"

"Every word. Look Bernice, I know Clarisse wants to do what's best for you, but my Brazilian viper venom is much more potent than her alfalfa extract. It's quite apparent that you are indeed a simple French farm girl, and as everyone knows, French farm girls have blood pressure as stubborn as they are."

For an instant you try to look offended, then you join him in laughter.

He gives you more samples of the ace inhibitor, but crosses out the brand name and in large block letters writes "Brazilian viper venom." As you leave the exam room, he says, "That's what I'd tell Clarisse. Tell her I surprised the hell out of you and started you on viper venom. The alfalfa didn't work, so Mother Nature forced us to use something stronger so you don't have a stroke— like your grandmother did, Bernice. Viper venom will save you from ending your days like your grandmother did."

You look over your shoulder and start to protest Doc's sledgehammer reminder that your grandmother had a stroke. It is a very painful memory. He looks at you over the top of his glasses and raises his eyebrows. It is a gesture that eloquently makes its point. It's not his blood pressure he's trying to control. It's yours. And your grandmother *did* spend the last four years of her life sitting in a wheelchair, spittle dripping from the corner of her mouth, unable to speak, and stripped of any vestige of personal freedom.

Part of you is offended that he would reach into

your past and use this haunting memory to make his point. Part of you is grateful that he knows you well enough to share in that memory and cares enough about your health to risk your displeasure.

You turn around and in a rush give him an affectionate hug. "You're the best, Doc. Even though there are times when you make me want to scream, you're the best."

As you leave, you are pleased to see that Doc's face is flushed with embarrassment.

Bernice's office visit is more ritual than melodrama. As a ritual, it has roots that extend back thousands of years—a healer using his innate, empathetic skills to enhance a patient's treatment. This is the primeval essence of a doctor/patient, healer/sufferer relationship. This is true holism. It is completely apart from the irrelevant marketing concept of a "natural" blood pressure medication.

For most of humanity's history, healers had to rely solely on their own intuitive skills. They used these skills to stimulate and supplement self-healing in their "patients." This was an intense, intimate interaction between sufferer and healer. It was an integral part of the structure of our earliest social units—the tribe or the clan. At some level, perhaps because it's been part of the human experience for tens of thousands of years, patients are aware of this. They seek this holistic approach to medical care. Because they rarely find it within the confines of traditional medicine, they look elsewhere. This is one of the ironies of Alternative Medicine. While many of its practitioners are charlatans, promoting worthless and ineffective treatments, sometimes the charlatans are blessed with the intuitive skills of a healer. Unfortunately, if these skills are used towards a self-serving end, patients suffer. It is also unfortunate that the circus-like aura of Alternative Medicine has so distorted true holism that traditional medicine has essentially ignored it.

Doc may not have burned incense or muttered chants as he treated Bernice, but he did use his intuitive skills to help her heal. In her case, healing meant dropping her blood pressure, and crucial to that healing was motivating her to take a medication that actually worked. Doc used the "parable of the carrot" to explain the ruse of "natural" treatments in a ritual not unlike a shaman evoking an age-old myth. He brought the ritual to a climax by reminding Bernice of her own family history, in a sense summoning the spirit of her grandmother. This is modern medical ritual at its best. Doc is obviously a gifted healer.

The difference between Doc's ritual and the melodrama of a slick salesperson on television hawking a sure fired weight loss pill is motive and honesty. Doc's motive was reducing Bernice's blood pressure. He pursued this goal with an intellectual honesty based upon logic and scientific observation. We know that lowering high blood pressure reduces the chances of strokes just as surely as our ancestors knew that an upwind approach to a wooly mammoth was a bad idea. Living with high blood pressure makes a person much more likely to have a stroke at a young age. Approaching a wooly mammoth from a position where the wind allows it to pick up a hunter's scent makes the hunter much more likely to catch one of those sixteen-foot tusks right in the groin.

The melodrama of the television infomercial is designed to sell a product in a manner both deceptive and misleading. Six thousand years ago, a shaman doing something similar would have claimed the power to drop a wooly mammoth with magic alone, magic that only he possessed. Of course, unlike today's infomercials that are repeated many times, the shaman's infomercial would have run but once. An irritated wooly mammoth enforces truth in advertising standards a bit more rigorously than they are now enforced.

The humanism and holism Doc used as he treated Bernice is infinitely more compatible with humanity's long-established traditions of healing than any vitamin with the word "natural" printed across its label.

Side effects

Anything a person puts in his or her mouth for the intended purpose of altering some physiologic function is a drug. It is a foreign substance ingested for a purpose other than either hydration or the intake of calories (energy). I don't care if the ingested substance is the bark of a willow tree or an aspirin tablet. The net effect in this case will be absorption of acetylsalicylic acid. But the willow bark might also have other effects, as anyone who has tried to swallow a piece of tree bark can tell you. That's why bottled willow bark is much less popular than Bayer Aspirin.

I will agree that eating a hunk of tree bark is more *primitive* than swallowing two aspirin, but there is absolutely nothing more natural about it. No other animals use willow bark for the purpose of analgesia or to prevent strokes. It is true that a number of North American Indians used it for medicinal purposes, but so did the Romans. *Natural* is an advertising concept, nothing more. But it's a damn good advertising concept. It's produced billions of dollars of profit from the sale of bottles that might as well have been empty.

I also do not deny that there are substances billed as natural that are biologically active. Saint John's Wart, for example, is an herb that traditionally has been used for a variety of medicinal purposes. Derived from the plant *Hypericum perforatum*, it has gained widespread use as a natural antidepressant. I have had many, many patients say to me, "I really don't want to take an antidepressant, Doctor. I don't need one of those. I'll just take some Saint John's Wart."

If the patients were significantly depressed, I would try to convince them to go on a synthetic antidepressant, not because Saint John's Wart doesn't help depression, but because it has a much longer lag in onset of action, it is much weaker, and the amount of antidepressant in a dose of Saint John's Wart varies all over the place. It is a ground-up extraction of a plant. The state of the plant at harvest, how much of the extraction is bud versus stem or flower, and whether or not the manufacturer uses filler all impact how much drug is in each dose. It's impossible for a patient taking Saint John's Wart to maintain a stable level of antidepressant.

Saint John's Wart works on the serotonin system just as does Prozac. When reports of side effects from the use of Saint John's Wart appeared in the medical literature and newspapers, I had a number of patients express absolute surprise. How could a natural medication have side effects identical to an unnatural creation like Prozac?

Saint John's Wart MUST have the same side effects as Prozac because it actually *works*. The only natural remedies that have no side effects are those that do *nothing* at all. For any medication to have a beneficial effect, it must be biologically active. This is true for roots, herbs, or a manufactured drug that costs five dollars a pill. Any substance that is biologically active will also have side effects—effects other than those for which the pill is taken—because a drug is spread throughout the body. Even if it does exactly what it is supposed to do in one organ, it reaches other organs where it may have an untoward effect. Diet pills that include ephedra, recently taken off the market, decrease appetite by centrally stimulating parts of the brain. Ephedra also gets to the cardiovascular system where it can produce fast heart rates and elevated blood pressure. It reaches other parts of the central nervous system and produces insomnia and anxiety. Saint John's

Wort can cause diarrhea, severe anxiety, dry mouth, interference with other drugs, impotence, and headache. So can Prozac.

Patients often protested that these remedies have been used for thousands of years, and asked, "How could they have those sorts of side effects?" My response to those arguments was simple. "True, and people have been having side effects for thousands of years." The first Emperor of China died because of a reaction to "natural" remedies designed to prolong his life (a rather ironic side effect).

The real side effects

I have not spent the time telling you about what I call snake oil just to anger those of you who think traditional medicine has it in for anyone not "in our club." Trust me—I know some of you are angry. I almost took this part out of the book because I did not want to have that effect.

I left it in because this is a prime example of how health care's lack of humanism impacts its effectiveness. The success of snake oil has nothing to do with the power of supplements or herbs to heal. Snake oil succeeds *in spite of* its inability to do much of anything because snake oil salesman understand the power of humanism. They understand that patients are people. They understand the value of trust. They use this understanding towards the end of selling lots of snake oil.

My patients spent hundreds of dollars on vitamins, forsaking the medications I prescribed, because they did not trust the traditional medical community. They trusted the high school student who worked part-time at the health food store more than they trusted me because the high school student had the time to talk to them, to *listen* to them. I promise you that I made a Herculean effort so that would not happen, but that requires more than Hercules' strength when one is seeing thirty patients a day.

Snake oil is a multi-billion dollar industry whose success is largely due to traditional medicine being disease oriented and not patient oriented. America's patients are working at cross purposes with its medical community because patients usually have no one to trust within that community.

We are just beginning to scientifically quantitate the power that healers have on people, to understand how human interaction in the form of compassion and understanding facilitates the inherent abilities of the body to heal itself.

Will this power set a broken bone? Hardly. But it *can* have an influence on the rate at which a bone heals. As we discussed when we reviewed a brief history of medicine, for thousands of years, this was all humanity had in its medical arsenal. Modern medicine has completely transformed that arsenal as it has come to understand the biologic mechanisms of disease, but patients are still human beings.

I'll bet you're really tired of hearing me say that.

The poker player learns that sometimes both science and common sense are wrong; that the bumblebee can fly; that, perhaps, one should never trust an expert; that there are more things in heaven and earth than are dreamt of by those with an academic bent.

-- David Mamet in "Things I Have Learned Playing Poker on the Hill," Writing in Restaurants (1986).

CHAPTER NINETEEN

MAKING IT SO

The most important specialty

I told you that the means of making medical humanism possible was simple but that its implementation was hard. The simple part begins with primary care. It must become the most important specialty in medicine. Instead of being a specialty area whose continued existence is in doubt, it must be the bedrock upon which the entire health care system is built.

Family practice, general medicine, and pediatrics must begin to draw those who have the unique blend of talents that produces healers—compassion, inherent joy dealing with people, intuition and compassion, and an intelligence that is centered on common sense, pattern recognition, and creativity. It must reward those with these talents in the same way it now rewards cardiologists and orthopedic surgeons.

If there are any cardiologists or orthopedic surgeons who have made it this far in the book, please—you'll hurt yourself jumping up and down like that. I know what I suggest is blasphemy and appears to be utterly self-serving given that I was a family practitioner. I admit it. You're right. Suggesting that the primary care docs should be rewarded the way you are is both blasphemous and self-serving.

It's also right. It's absolutely necessary for health care to progress beyond where it is.

For all the reasons we have discussed, primary care must be the engine that drives medical care. As treatments of diseases become more complicated, as the population ages, there must be a force that coordinates the multiple modes of care prescribed by multiple specialists and sub-specialists. Without that force, patients' lives will be utter chaos.

Patients need their own doctors. They need someone they trust. That can only come in the setting of primary care because primary care is designed to *fill that role*. That is its purpose. What is needed is the recognition that this role is important. Right now, the patient as a whole is being ignored, and that is why patients are unhappy. That is why their care is chaotic.

The bedrock

I did not use the word "bedrock" because of artistic license. I used it because recognizing primary care's importance will create exactly that—a base upon which health care can build. If health care's supply side is to be molded by the forces of free market competition, there must be some stable structure around which that evolution takes place. Primary care offers that structure.

When I use the term primary care, I am not just referring to family practitioners, pediatricians, general internists, gerontologists, and ob-gyns. I am also referring to the mid-level providers (physician assistants and nurse practitioners) and the ancillary services that should work in concert with these physicians.

The face of primary care will change as it evolves along with the rest of the medical community. It will never be a specialty of ol' Doc Johnsons, laboring in the isolation of solo practice. Even the most devoted physicians loathe the thought of a life of being on call 24/7, 365 days a year. That sort of existence quickly burns away a physician's humanism. That sort of existence destroys marriages, creates alcoholics and drug addicts, and dumps doctors into premature graves.

Primary care will probably emerge from its evolution as teams of professionals. Doctors, mid-level providers (MLPs), social workers, and visiting nurses may well form functional units that deal with the challenges of a rapidly aging population. These units will be geared towards coordinating not only complicated and interacting treatment regimens but also the various services necessary to assist this aging population as it seeks continued independence. Facilitating independence for people who have survived into their eighth and ninth decades is becoming one of health care's greatest challenges.

It is ironic that disease oriented medicine's single minded objective of prolonging life utterly failed to consider the implications of that objective. Unless humanism is applied through the conduit of primary care, longevity will become a national tragedy rather than a gift of modern medicine. Longevity will represent a prison sentence served in the unnatural environment of nursing homes.

As the wave of aging Americans becomes a tsunami, it will fall upon health care to mitigate the impact of this potential social catastrophe. The cost, impact on productivity, and strain on social services can be held to a minimum, but only by a health system that is superbly integrated.

Only an empowered primary care can supply that integration.

The process of evoltion

With a RIS in place, reimbursement for primary care can be brought to levels commensurate with other specialties. This would be the first example of DDHC applying pressure to create appropriate change. The supply side will then respond to that pressure. It will move towards meeting an increased demand for a service, that demand being reflected in its increased price. But the way the supply side *specifically* responds will be determined by the "wisdom of the marketplace" not by the decree of some massive bureaucratic entity.

There will be increased incentive for medical students to become primary care doctors and primary care doctors already in practice to remain in practice because practicing primary care will no longer require financial sacrifice in addition to commitment.

This is only reasonable. Why should a professional who has competed against the stiffest competition in order to earn a place in medical school, withstood the rigors of 80 hour work weeks during training, and sacrificed a decade of his or her earning potential also be asked to then work harder and earn less for the rest of a professional life? Is there *any* reason to believe that this situation will attract the highest quality doctors? It's amazing that it attracts any doctors at all.

Once primary care is financially situated as an attractive specialty, competition will develop to fill its ranks, the same way there is now competition to fill the ranks of specialties like dermatology. For America, the difference is that the country desperately needs more general internists, pediatricians, and family physicians. Its need for more dermatologists is much less acute.

It will take time for the shortage of primary care doctors to be resolved. How this shortage is met, i.e. what the mix of MLPs to physicians is, and the configuration delivery systems, will also require some time. But the system will have the freedom

345

to find a solution to these questions. Various solutions will compete with regard to quality and efficiency, and at some point, equilibrium will be established.

This is the advantage of DDHC. It avoids the necessity for creation of a uniform model of health care delivery as is the case with the so called single party payer because the supply side is left untouched. Any change in America's health care system that involves a complete overhaul of all doctors and all hospitals and all other parts of medical care will have devastating consequences. Transition to a single, unified system would be a nightmare. Further, at present, there isn't even a successful model to adopt. If there were, Great Britain and Canada would not be redesigning theirs.   There are a number of delivery systems that have been proposed as being the most efficient or the highest quality. Perhaps one of these systems is clever enough that it blends all the forces of efficiency, reward for effort, and humanism and integrates them into an ideal way to deliver care. But committing to *one* such system and converting all of America to that system is *impossible.* This makes the elimination of the replacement of the medical insurance companies with a RIS look like a walk in the park.

And humanism?

Let's review what we have accomplished so far. We have created the ultimate consumer by replacing all of America's various funding sources with a RIS. We have also saved billions of dollars by doing so—just in efficiency and elimination of redundant bureaucracy. Finally, we have strengthened primary care by rewarding it appropriately.

If you think about it, that doesn't really sound all that complicated. Yes, it's dramatic—eliminating an industry that handles a trillion dollars of cash flow is certainly dramatic. But that's not actually complicated. A RIS would handle insurance claims the same way they are now handled, just more efficiently.

The increased reimbursement for primary care would be controversial. Dollars for primary care would have to come from somewhere, and it is conceivable there may need to be reduced reimbursement for other specialties. That would potentially create some embittered debate, but a RIS is designed to handle such debate in an efficient and rational fashion. Whatever decisions to be made would be based upon balanced input and with high quality, low cost care the only goal. But the change in reimbursement itself would be easy to implement.

So, how does any of this promote humanism?

Nurses, physical therapists, pharmacists, occupational therapists, emergency room clerks, x-ray technicians, receptionists, scheduling clerks, and medical assistants have far more day-to-day patient contact than doctors do. Thus, these medical professionals play a crucial role in whether or not a health care system is humanistic or impersonal. How does any of what we have created address their part of the equation?

The answer is the "magic of the marketplace." This is

why DDHC is the only system that utilizes America's strength. A RIS truly *is* an ultimate consumer. It can establish humanism as one of its "rational" means of making economic decisions. It has the capacity to apply a cuantitative measure to the qualitative entity of humanism.

A RIS offers a viable means of collecting patient feedback. Now, if patients feel that they were hustled through emergency rooms, their only recourse is to complain to whoever runs that emergency room or to their insurance company or to their personal physicians.

Big deal.

Those complaints have no leverage. The company running the emergency room is not competing to please patients. Emergency rooms are so overwhelmed that they are simply trying to function while knocking off the fewest number of people. The insurance company doesn't care. Their goal is to negotiate the lowest unit cost so there will be money left over to distribute to their shareholders. A patient's doctor? Leverage? Hardly.

A patient who feels like a number or worse has no ability to influence anything.

But with a RIS, humanism can be factored into levels of reimbursement. Patients seen emergency rooms can give immediate feedback about their experiences. That feedback can be factored into levels of reimbursement for every emergency room. Every patient thus becomes empowered because a RIS is empowered.

Those emergency rooms that have admitting clerks who are trained to recognize patients who are overwhelmed and have the means and time to address this state with understanding and compassion will be rewarded—financially. Those that do not will be penalized. Thus are the ways of the market, and thus would humanism become part of health care.

Mercenary? Yes. But asking altruism to compete against profit does not promote altruism. It hinders it. Profit and altruism, financial advantage and human compassion should be aligned, not opposed. If compassion is penalized, it will be found *in spite of* and not *because of* a health care system. If it is rewarded, it will flourish.

The same feedback system can be applied to any health care experience. A RIS will have the ultimate tool to gauge and reward health care—data. Every health care encounter will leave a piece of data that will flow through a single source. This sort of data can be used to accomplish countless tasks—even the promotion of a caring and humanistic health care system.

Medical Metamorphosis

.

EPILOGUE

There are very few issues that directly benefit a person while still benefiting society as a whole. Usually, issues are a zero sum proposition—an individual sacrifices for the good of the whole. In fact, the meaning of "altruism" includes exactly that reality.

What we have been talking about is an exception to that rule. It is an issue that is *not* zero sum. Improving the health care system as we have suggested does not just benefit those Americans who are at present without health care. It benefits *all* Americans because it offers a means of controlling health care costs, increasing health care quality, and facilitating the creativity and strength of America's businesses and industries.

Critics of this book may dismiss its thesis because it lacks the detail that will be needed to implement its programs. I would imagine the phrase, "all belt buckle and no horse" may well be used in this context.

My response to that criticism is to note that the Clinton health care proposal exceeded 1,300 pages. There are those who feel that simply the length of the proposal doomed it to fail. No one with the power to implement change will read a 1,300 page proposal. The success that legislators enjoy hiding pork barrel additions in the federal budget is ample proof of that fact.

As this book goes to its publisher, presidential candidates are touting their proposals for America's health care system. Health care has finally made it onto the stage of national priorities.

Unfortunately, *none* of these proposals will create the medical metamorphosis we have been discussing. None of them address the simple realities that have created a dysfunctional system, and none of them are fashioned from the courage and honesty necessary to suggest that the "players" must be confronted, regardless of the political implications of that confrontation.

The political sages scoff at any health care policy more dramatic than what has already been suggested. They call it "suicidal" to propose the sort of change represented by DDHC and a RIS. But America cannot continue to allow a small number of businesses and industries to determine the future of an issue that influences *every other business and industry and every American as well.* America cannot continue to pay heed to the voices of political and philosophical fanaticism while *every year, 100,000 American citizens die needless deaths.*

Those who suffer must be the catalyst for change.

I feel a bit like Smokie the Bear, the old icon of forest fire prevention. He pointed his finger from countless posters and admonished America to realize, "Only *you* can prevent forest fires."

I'm pointing my finger and saying, "Only *you* can bring about health care change." Those running things will simply not endure the risk until forced.

How do you apply the force?

First, you must realize that there is a way to institute a national health care program without damaging what's not already damaged. DDHC leaves the hospitals, doctors, and pharmaceutical companies completely intact.

Second, you must realize that a RIS will not be a new government bureaucracy. It will be entirely separate from the government, and it will replace *scores* of inefficient and unnecessary bureaucracies whose goal was profit first, quality of care second.

Third, you must realize that any efficient, effective health care system must be anchored by a strong, viable base of primary care. Without this anchor, health care will flounder in the storm of new technology and eventually capsize under the waves of an aging population.

Finally, you must spread the word—share your understanding. Right now, a political candidate can propose a totally dysfunctional health care system designed by economists and not be questioned by anyone other than economists from the opposite political party. Average Americans are lost in the fog of complicated nonsense. Burn away that fog by sharing what we have discussed.

Medical metamorphosis is possible. We can turn the snarling snake on the caduceus—the symbol of commerce and theft—into a butterfly of compassionate care, an alleviation of pain and suffering, and the restoration of the whole.

We can. We really can.

Medical Metamorphosis

www.ingramcontent.com/pod-product-compliance
Lightning Source LLC
Chambersburg PA
CBHW031818170526
45157CB00001B/107